T0291112

Conventions for calculating linear thermal transmittance and temperature factors

Tim Ward, Graeme Hannah and Chris Sanders*

* Glasgow Caledonian University

The research and writing for this publication has been funded by BRE Trust, the largest UK charity dedicated specifically to research and education in the built environment. BRE Trust uses the profits made by its trading companies to fund new research and education programmes that advance knowledge, innovation and communication for public benefit.

BRE Trust is a company limited by guarantee, registered in England and Wales (no. 3282856) and registered as a charity in England (no. 1092193) and in Scotland (no. SC039320). Registered office: Bucknalls Lane, Garston, Watford, Herts WD25 9XX
Tel: +44 (0) 333 321 8811
Email: secretary@bretrust.co.uk
www.bretrust.org.uk

IHS (NYSE: IHS) is the leading source of information, insight and analytics in critical areas that shape today's business landscape. Businesses and governments in more than 165 countries around the globe rely on the comprehensive content, expert independent analysis and flexible delivery methods of IHS to make high-impact decisions and develop strategies with speed and confidence. IHS is the exclusive publisher of BRE publications.

IHS Global Ltd is a private limited company registered in England and Wales (no. 00788737).
Registered office: The Capitol Building, Oldbury, Bracknell, Berkshire RG12 8FZ. www.ihs.com

BRE publications are available from www.brebookshop.com
or
IHS BRE Press
The Capitol Building
Oldbury
Bracknell
Berkshire RG12 8FZ
Tel: +44 (0) 1344 328038
Fax: +44 (0) 1344 328005
Email: brepress@ihs.com

Printed using FSC or PEFC material from sustainable forests.

BR 497
First published 2007
Second edition 2016
ISBN 978-1-84806-440-9

Any third-party URLs are given for information and reference purposes only and BRE and IHS do not control or warrant the accuracy, relevance, availability, timeliness or completeness of the information contained on any third-party website. Inclusion of any third-party details or website is not intended to reflect their importance, nor is it intended to endorse any views expressed, products or services offered, nor the companies or organisations in question.

Any views expressed in this publication are not necessarily those of BRE or IHS. BRE and IHS have made every effort to ensure that the information and guidance in this publication were accurate when published, but can take no responsibility for the subsequent use of this information, nor for any errors or omissions it may contain. To the extent permitted by law, BRE and IHS shall not be liable for any loss, damage or expense incurred by reliance on the information or any statement contained herein.

The first edition of this report was produced as part of the research programme of the Sustainable Buildings Division of the Department for Communities and Local Government. This revision was produced as part of the research programme of the BRE Trust.

Any updates to the guidance given in this publication will be posted at www.bre.co.uk/br497updates

Contents

Cont'd . . .

Abbreviations and notation

Abbreviations

dpc	damp proof course
dpm	damp proof membrane
low-e	low emissivity
P/A	perimeter divided by area ratio
1D	one dimension/one-dimensional
2D	two dimensions/two-dimensional
3D	three dimensions/three-dimensional

Notation

A	area
b	width
β	angle of slope
χ	point thermal transmittance
d	thickness
e	external environment
f	temperature factor
h	height
h_{se}	heat transfer coefficient of external surface
h_{si}	heat transfer coefficient of inside surface
H	heat transfer factor
i	internal environment
L	thermal coupling coefficient
ℓ	length in metres over which U applies
n	number of units
Q	total heat flow
R_{se}	thermal resistance of external surface
R_{si}	thermal resistance of inside surface
T	temperature
U	U-value
w	width
Ψ	linear thermal transmittance

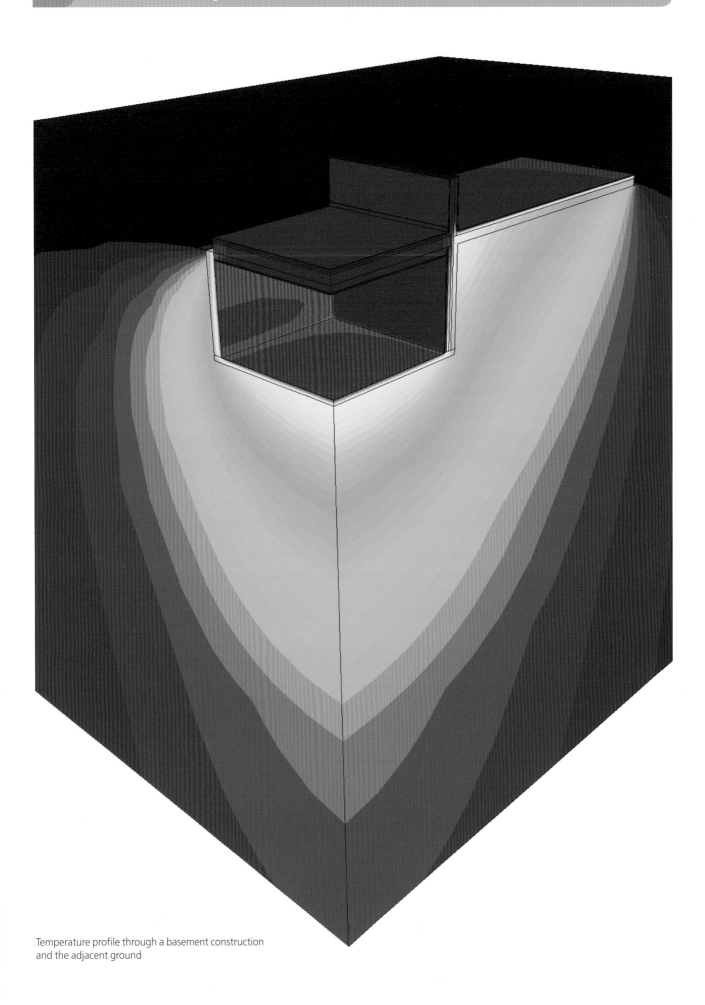

Temperature profile through a basement construction
and the adjacent ground

1 Introduction

Global warming, with the need to limit CO_2 emissions into the atmosphere, is the principal driver for conserving fuel and power in buildings. Confirmation that this is very much part of the UK Government's agenda is found in the successive changes to Part L of the Building Regulations in England[1] and Wales[2], with similar changes to the equivalent sections in Scotland[3] and Northern Ireland[4].

The introduction of more highly insulated buildings, that has resulted from the need to save energy, has also led to the need for more sophisticated methods for calculating heat loss and surface temperatures than were previously felt to be adequate. Two changes are particularly important:

- While U-values of the building fabric could previously be calculated by assuming that an element was made up of a series of parallel layers, each with uniform thermal resistance, it is now recognised that features such as mortar joints, timber studs or the metal spacers in built-up roofs cause thermal bridging of the insulation layer(s) and so contribute significantly to the heat loss. A more detailed calculation method for U-values, as defined in BS EN ISO 6946:2007[5], has been introduced to take account of these repeating thermal bridges.
- It has also been recognised that thermal bridging at the junctions between the various plane building elements (walls, roofs and floors) of a building and those around openings in walls and roofs can add significantly to the fabric heat loss. The higher heat flows that occur, because of complex geometries or the use of materials with a high thermal conductivity, also cause localised reduction in the internal surface temperatures, which in turn can lead to surface condensation and mould problems.

Although various simplified calculation methods have been developed to take account of the effects of thermal bridging in certain situations, two- or three- dimensional heat flow calculations continue to be required for *some* U-value and for *most* (non-repeating) thermal bridge calculations. These calculations of two- or three-dimensional heat flow require the use of numerical modelling software. Several packages are available but, whereas most software packages themselves are validated as being able to produce correct and consistent results, many important decisions are left to the user regarding the input to the modelling software and the determination of certain quantities from the output of the software, both of which can have a significant effect on the results.

This BRE guide (BR 497) gives the information needed to carry out these calculations so that different users of the same software package and users of different software packages can obtain consistent results. However, *before using the conventions* given in BR 497 it is important for the numerical modeller to demonstrate that the numerical modelling software used can model the validation examples in BS EN ISO 10211:2007[6] with results that agree with the stated values of temperature and heat flow within the tolerance indicated in the standard for each appropriate validation example.

BR 497 has been prepared to complement the outline methodology for the treatment of thermal bridges given in BRE Information Paper IP 1/06[7]. It can be used by assessors who wish to undertake numerical modelling calculations to determine the thermal performance of junctions. It is referenced in the relevant government policy documents operating in England, Scotland, Wales, Northern Ireland and the Republic of Ireland.

2 Numerical modelling

2.1 Two- or three-dimensional modelling

Thermal bridges occur within the building fabric where, because of the geometry or the presence of high conductivity materials, heat flows are two- or three- dimensional (2D or 3D). For many situations, simple calculations are no longer sufficient to determine thermal performance correctly and it is necessary to analyse the construction using numerical modelling. A number of numerical modelling software packages are available which specify the geometry, the materials and the boundary conditions of the model in two or three dimensions as appropriate.

A linear thermal bridge is an essentially 2D concept and, since it is much easier and quicker to specify a detail in 2D rather than 3D, a 2D model should be used wherever it is adequate. However, as discussed later in section 3.1.3, when the plane elements of a structure being analysed are non-uniform in the third dimension, ie where the plane (flanking) elements contain repeating thermal bridging, it may be necessary to use a 3D model of the junction in order to properly include the influence of this repeating thermal bridging on the calculated linear thermal transmittance, Ψ, of the linear thermal bridge.

Where the temperature factor, f, is of concern, the lowest surface temperatures may occur where two or more linear thermal bridges meet. In this case, a 3D model is often required. Indeed for the corner of a ground floor (see section 3.2.1) a 3D model *must* be used.

Modelling in 3D is also necessary to assess the heat loss and lowest inside surface temperature in the case of individual penetrations through the plane elements (see section 5).

2.2 Specification of the numerical model

2.2.1 General principles

The fabric heat loss can be divided into:

*   heat loss through the plane elements, represented by their U-values
*   heat loss through the (non-repeating) thermal bridges, which are generally the junctions between the plane building elements or around openings, represented by Ψ-values.

In any detail, the lowest internal surface temperature and therefore the temperature factor may be associated either with one of the plane areas or with the thermal bridge.

When a structure is being analysed it is important to decide which features affect the heat flow through the plane areas and should be assigned to the U-value, and which affect the heat

flow through the thermal bridge and should be assigned to the Ψ-value; otherwise, there is a danger that some features will be included in both and their effect on the heat loss counted twice.

The requirements for model development are discussed further in clause 5 of BS EN ISO 10211:2007[6].

2.2.2 Extent of the model

When a model is being constructed, the flanking elements of the model (the plane areas adjacent to the thermal bridge) should be taken to at least 1 metre or three times the thickness of the flanking element, whichever is greater, away from the thermal bridge, or up to a plane of symmetry in the case of repeating features. If there is any uncertainty as to whether the model extends far enough for a particular flanking element (or elements), the internal (or external) surface temperature at the adiabatic edge of the particular flanking element(s) should be noted, the model extended by at least the thickness of the flanking element, and the surface temperature at the 'new' adiabatic edge of the flanking element recalculated.

If the difference in temperature factor (between successive models) is no more than 0.005[1], the smaller model is adequate; otherwise, this process of extending the model should be repeated until the condition for that particular flanking element (or elements) is met.

The dimensions to use for ground floors are given in clause 5.2.4 of BS EN ISO 10211 : 2007[6], where the key dimension that determines the overall size of the model is the breadth of the floor. In the case of a 3D model (used for calculating the temperatures in the corner of the floor), consider the floor to be square, where the breadth (and length) of the floor, B, is taken to be 8 m. The model should then cover a quarter of the floor where the distance from each edge of the floor to the opposite adiabatic boundary is 4 m. In the case of a 2D model of the floor the 'breadth' of the floor is the *characteristic dimension*, B' (see clause 5.2.4 of BS EN ISO 10211:2007), where this is taken to be 8 m. The distance from the edge of the floor in the 2D model to the opposite adiabatic boundary is then 4 m. For the calculation of the U-value of the ground floor, B'= 8 m corresponds to a perimeter/area ratio (P/A) of 0.25.

2.2.3 Adjacent thermal bridges

There are many instances where one thermal bridge may be close to another. For example, a lintel will often be less than half a metre from the wall/roof or wall/intermediate floor junction. If they are sufficiently close for the heat flow through one to affect the heat flow through the other, analysing them separately to derive two separate Ψ-values will overestimate the heat loss.

---

[1] For an internal temperature of 20 °C and an external temperature of 0 °C this difference in temperature factor is equivalent to a difference in temperature of 0.1 °C.

It is also possible that the minimum surface temperature, and therefore the temperature factor, will be lower on the combined detail than it is on either of the individual ones. If two junctions are less than the thickness of the building element apart they should be included in the same model, otherwise they should be treated separately.

If they are to be included in the same model, consideration needs to be given to the lengths over which the separate junctions apply, in order that the heat loss contribution from these two junctions is correct. If the length over which each separate junction applies is the same, then that length is used with the Ψ-value of the combined junction. However, if these lengths are different then the combined Ψ-value needs to be divided such that the portion assigned to the longer junction length is equal to the Ψ-value of that longer junction, where this is calculated separately, ie not combined with the shorter junction. The remaining portion of the combined Ψ-value is assigned to the shorter length junction. These divided Ψ-values then apply over the respective lengths of the two junctions. To illustrate, consider an eaves and lintel combined junction.

The process would be:

1   Model the eaves detail in isolation to give Ψ_{eaves} (This is required in any case for the lengths of the eaves where its junction is not combined with any other junction.)

2   Model the combined junction to give $\Psi_{eaves+lintel}$

3   Subtract Ψ_{eaves} from $\Psi_{eaves+lintel}$ to give $\Psi_{lintel\ (combined)}$

4   Obtain the heat loss associated with Ψ_{eaves} and $\Psi_{lintel\ (combined)}$ by multiplying by the length over which each applies.

2.2.4 Non-rectangular elements

Some modelling software packages can deal only with rectangular shapes (ie shapes that can be represented by rectangles with sides (or in 3D, faces) parallel to the X, Y and Z axes). When these packages are used to analyse non-rectangular elements, the sloping parts in the geometry of the detail have to be approximated by a series of steps. The size and the number of steps should be determined by considering the angle of the sloping part and the thickness of any thin layer along the slope, if present. Thin layers are defined as layers whose thickness is no greater than 4 mm. Using the notation shown in Figure 1, β is the angle of the slope and $y/x = \tan(\beta)$. If $\tan(\beta) > 1$, ie $\beta > 45°$, then $x < y$, otherwise $x > y$. The smaller of x or y must be no greater than 2.5 × the thickness, d, of the sloping thin layer. For any sloping part of the geometry that is not part of a thin layer, the smaller of x or y must be no greater than 10 mm.

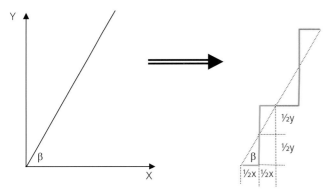

Figure 1: Stepping of slope for rectangular modelling software

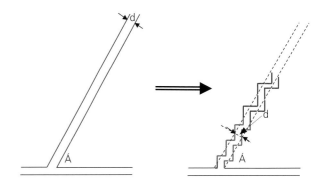

Figure 2: Diagram illustrates the stepping of a sloping layer

Within these maximum limits the step size need not be the same for all steps and can be larger or smaller than an adjacent step. Indeed this is likely to be needed in order to have a whole number of completed steps between the start and end of the slope being approximated, ie no partial steps. Figure 2 shows two adjacent and differently sized steps – one marked in green and one in blue. The correct intersection of each step with the line of the slope being approximated should produce two equal triangles (one above and one below the line).

The stepping arrangement shown is for any thickness of sloping layer. (Only one end of the sloping layer is shown.) To determine the stepping arrangement, one sloping line of the thin layer is considered first and its steps are formed following the rules given above. In the illustration shown in Figure 2 the lower line is formed first. The first three steps are shown in green and the remaining two (larger) steps are shown in blue. The second (upper) line (apart from the end step shown) is an exact image of the first but moved a distance $d \cdot \cos\beta$ to the left and $d \cdot \sin\beta$ upwards. Depending on the precise connection of the sloping part with the non-sloping parts of the detail being modelled, the end(s) of one or more of the stepped lines along the slope may need to be extended (or shortened) to meet with the rest of the rectangular model. In the illustration given in Figure 2 this connection is shown (in this case extended) with red lines.

2.2.5 Large spaces with an intermediate temperature between the internal and external environment

Some details may contain large ventilated spaces between the internal and external environments. These include loft spaces in pitched roofs and air spaces below suspended floors and are considered to have a temperature intermediate between the internal and external temperatures.

2.2.5.1 Air space below a suspended ground floor
In the case of an air space below a suspended ground floor, the method of determining the temperature of the air space is via the heat balance as described in Annex E of BS EN ISO 13370:2007[8]. This temperature is one of the calculated outputs from the BRE U-value calculator[9].

2.2.5.2 Air space in a cold loft
In the case of a cold loft space its temperature depends on the heat balance to and from the cold loft space, where the heat into the cold loft space is then lost via transmission through the remaining roof structure and via ventilation of the cold loft space. In the case where the ceiling is uninsulated or poorly

insulated, the loft-space temperature is at most about +1 °C above the external temperature. For well insulated lofts with full insulation between (or between and over) the ceiling joists, the loft space temperature is closer still to that of the external air temperature. The reduction in Ψ-value for an increase in loft temperature of 1 °C above the external temperature is at most approximately 0.002 W/m·K. The temperature of a cold loft space can therefore be taken to be the same as that of the external air, irrespective of the level of insulation to the ceiling. The resulting Ψ-value is only slightly pessimistic compared with using a precise heat-balance temperature for the cold loft space. For example, a 'correct' Ψ-value of, say, 0.12 W/m·K for an eaves junction with an uninsulated or poorly insulated ceiling would become approximately 0.122 W/m·K and for a well insulated ceiling would become approximately 0.121 W/m·K.

2.2.6 Substitution of window and door frames with adiabatic boundary layers

To avoid repeating the assessment of junctions around openings for each different frame type (or in the situation where the frame type is unknown), the connection of the frame with the opening should be replaced with an adiabatic boundary layer, ie one across which there is no heat transfer. Using such an adiabatic substitute for the frame allows different opening surrounds to be more readily compared. In some instances, however, the temperature factor so determined can be a little optimistic. Although this approximation is acceptable for regulatory purposes, where full details of the frame of the opening are known and a more accurate minimum inside surface temperature (and hence minimum temperature factor) is wanted, then a second model can be made which now includes the known frame. Importantly, this second model is for *temperature assessment only* and should not be used to determine the Ψ-value of the junction detail. This *must* be determined from a model that substitutes for the frame with an adiabatic boundary.

2.3 Thermal conductivities of materials

The detailed guidance in *Conventions for U-value calculations* (BRE Report BR 443)[10] should be followed for the thermal conductivity values to use when modelling constructions. Besides manufacturers' declared thermal conductivity values for their products (mainly insulation materials), manufacturers' literature also provides information on specific products. The thermal conductivities used in calculations, together with their source (if known), should always be listed with the results.

2.3.1 Perforated metal plates

Metal lintels and other metal components are sometimes perforated to reduce heat flow through them. To avoid the problem of trying to model the perforations as part of the much larger model, the perforated metal plate can be separately modelled to obtain an effective thermal conductivity. The metal plate is then included in the large model as a solid material with that effective thermal conductivity. The perforated plate is modelled in 2D with constant temperatures on opposite edges of the model. If the software cannot set all the surface nodes to a specified temperature, eg 20 °C inside and 0 °C outside, then the surface resistance (m²K/W) or surface conductance

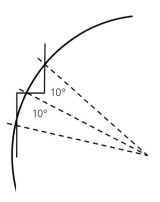

Figure 3: Stepping arrangement for a curve

(W/m²K) should be set to around 10 $^{-7}$ or 10 ^{7}, respectively. If the modelling software is restricted to larger values of surface resistance (or to smaller values of surface conductance) than these, this may lead to significant computational errors.

When using these demanding surface conditions, the difference between the total heat flow into the warm surface and that out of the cold surface should be checked to ensure it is no greater than 1%[2]. The orientation of the slots (with respect to the principal heat flow direction) must be the same as that of the plate in the detail. If the slots repeat in a regular pattern only one instance of the repeat need be modelled, ie the symmetry lines of the repeat form the four sides of the model of the perforated plate, otherwise the whole plate needs to be modelled. If there are any sloping edges to the slots that need to be approximated in a rectangular system, then this should be done by using the same rules as in section 2.2.4. If there are curved edges to the slots then these should be approximated with an arrangement of steps such that successive vertical and horizontal lines forming these steps intersect with the curve being approximated, with successive intersections of curve and step at a separation on the curve that is no more than 10° apart, as shown in Figure 3.

This technique of modelling repeating elements and obtaining an equivalent thermal conductivity for use in the larger model may be used more generally to construct 'quasi-homogeneous' layers to simplify modelling (see 5.3.3 of BS EN ISO 10211:2007[6]).

2.4 Treatment of air spaces

Heat transmission through air cavities within a structure is by a complex mixture of conduction, convection and radiation, and depends on the dimensions and orientation of the cavity, the emissivity of the surfaces and the degree to which the cavity is ventilated to the external air.

Nevertheless, because the radiation and convection heat transfer across them is approximately proportional to the temperature difference between the bounding surfaces, an air space can reasonably be treated as a material with a specified thermal conductivity. For the purposes of modelling according

---

2   If the error is larger than this then the calculation of an equivalent thermal conductivity for the perforated plate should be abandoned and the perforated plate included in the larger model using its normal thermal conductivity.

to these conventions, an air space is assigned an equivalent thermal conductivity, obtained by dividing its thickness in the principal heat flow direction by its thermal resistance. In the model the principal heat flow direction for any air space is restricted to one of the three orthogonal directions in the model. The principal heat flow direction is usually obvious for air spaces whose height and width are much larger than their thickness, such as the major cavities in wall constructions. For divided air spaces (see sections 2.4.2 and 2.4.3) the principal heat flow direction is sometimes less obvious.

The emissivity of most building materials is high and for calculation purposes should be taken to be 0.9. However, materials with a polished metal surface, such as aluminium foil, have lower emissivities of 0.2 or less. If either or both of the surfaces of a cavity (or opposite faces of a divided air space) have low emissivity, the radiative component of heat transfer across the cavity or air space is reduced and so its thermal resistance is higher, ie its effective thermal conductivity is less, compared with that of normal high emissivity cavities or air spaces. It should be assumed that the emissivity of all materials is high unless there is acceptable independent test evidence[3] to the contrary. See also section 4.8.2 of BRE Report BR 443[10].

2.4.1 Air space resistances

2.4.1.1   Unventilated cavities
For all unventilated air spaces (excluding those within the frames of windows or doors, which are dealt with in accordance with BS EN ISO 10077-2:2012[11] the mean temperature of the air space is taken to be 10 °C with the temperature difference across it taken to be 5 °C. Use the formula in Annex B of BS EN ISO 6946:2007[5] to calculate the air space resistances to use when modelling a construction. Where appropriate, see also section 2.4.1.2 below.

2.4.1.2   Ventilated cavities
If a cavity within the building structure is affected by air flow from the external environment its thermal properties will be modified. BS EN ISO 6946:2007[5] distinguishes between three cases.

1 *Minimal ventilated cavity*
If there is no insulation between a cavity and the external environment and small openings to the external environment, it can be considered as an unventilated air layer, if these openings are not arranged so as to permit air flow through the layer and they do not exceed:
 – 500 mm² per m length for vertical air layers
 – 500 mm² per m² of surface area for horizontal air layers.

2 *Slightly ventilated cavity*
A slightly ventilated cavity is one in which there is provision for limited air flow through it from the external environment by openings of area, A_v, within the following ranges:
 – > 500 mm² but < 1500 mm² per m length for vertical air layers
 – > 500 mm² but < 1500 mm² per m² of surface area for horizontal air layers.

3 *Well ventilated cavity*
A well ventilated cavity is one for which the openings between the air layer and the external environment are equal to or exceed:

 – 1500 mm² per m length for vertical air layers
 – 1500 mm² per m² of surface area for horizontal air layers.

For cases 1 and 2, the air space resistance of these ventilated cavities should be treated as unventilated cavities and the air space resistance calculated as in section 2.4.1.1. For case 3, the well ventilated cavity, the total thermal resistance of a building component containing the well ventilated cavity is obtained by disregarding the thermal resistance of the cavity and all other layers between the cavity and the external environment, but should include an 'external' surface resistance (corresponding to that for still air and using the appropriate emissivity[4] for the material of the 'external' surface). In the model, the same layers should of course be disregarded and the appropriate 'external' surface boundary condition used.

2.4.2 Regular divided air spaces

Regular divided air spaces are air spaces in the model where the thickness, width (and in 3D models, the length) are orthogonal and where the ratio of the width of the air space to its thickness in the principal heat flow direction is no greater than 10. In a 3D model the width of the air space is the smaller of the two remaining dimensions after deciding on the thickness in the principal heat flow direction. The thermal resistance is then calculated using the equations in Annex B of BS EN ISO 6946:2006[5] for divided air spaces, where the mean temperature of the air space is taken to be 10 °C and the temperature difference across it is taken as 5 °C and, where appropriate, taking into account the presence of any low emissivity surfaces on either of the opposite faces (at right angles to the principal heat flow direction) of the air space. The equivalent thermal conductivity is obtained by dividing the thickness of the air space in the principal heat flow direction by its calculated thermal resistance.

2.4.3 Irregular divided air spaces

Where a divided air space is irregular or has a complex shape, ie is not a simple rectangle in 2D (or cuboid in 3D), for the purposes of determining its thermal conductivity the shape of the air space is transformed into a rectangle (or cuboid) of the same area (or volume) as the original shape and with the same aspect ratios as the original shape. The aspect ratios are the ratios of the maximum dimensions of the shape, but restricted to the three orthogonal axes of the model.

Figure 4 gives an example (a) of such an air space in 2D and illustrates the transformation (b) to the appropriate rectangle. Figure 4(a) is the shape from the drawings and is not necessarily the shape of this air space in the model, where, for example, this shape has had to be approximated (see section 2.2.4). In (a), B is the maximum dimension in the Y direction, D is the maximum dimension in the X direction and A is the area of the complex

---

[3] In the absence of specific evidence, lower emissivities can be used for the generic material provided it is from a reliable published source.

[4] If both surfaces to the fully ventilated cavity are low emissivity, then when calculating the 'external' surface resistance in accordance with Annex A of BS EN ISO 6946:2007[5], the emissivity to use is the intersurface emittance determined from equation B3 of Annex B of BS EN ISO 6946:2007. If there is only one low emissivity surface and the other is taken to have normal emissivity (0.9) then the emissivity to use is simply that of the low emissivity surface.

(a)

(b)

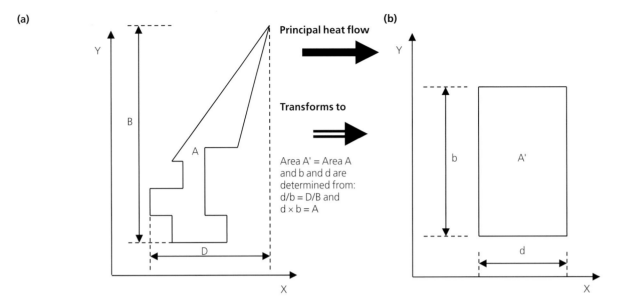

Figure 4: Transformation of a complex air space from (a) to (b)

shape, where D, B and A are from the drawings (or the model, if appropriate). The principal heat flow is in direction X and the complex air space is transformed to the simple rectangle whose area, A', is equal to that of the complex shape (A) with thickness d in the direction of the principal heat flow and width, b, perpendicular to the principal heat flow direction and where the ratio of d/b is equal to D/B. As for regular divided air spaces, the thermal resistance of the transformed air space is then calculated using the equations in Annex B of BS EN ISO 6946 for divided air spaces. *Note:* the shape of the actual air space in the model is unchanged. The transformation as described is only for determining the equivalent thermal conductivity to use for the air space in the model.

2.4.4 Narrow extended air spaces

Narrow extended air spaces, within which the magnitude and direction of the heat flow varies greatly, require special consideration. A narrow extended air space in a 2D modelled

junction is one in which the length of the extended air space (larger dimension) is in the direction of the principal heat flow (which is towards the junction) and where the ratio of the larger dimension to the width of the air space (smaller dimension) is greater than 10. Consider an air space within the construction of an intermediate floor as shown in cross-section in Figure 5, where the air space extends across the width of the floor and close to the opposite edges of the floor. This is defined as an extended air space where B/d is greater than 10.

The upper and lower rooms are at the same temperature and, consequently, at the centre of the floor there is little or no heat flow entering the floor from the top or bottom, while as the edge of the floor is approached the heat flow into the floor increases and is progressively channelled towards the edge of the floor. Such an air space requires (in the model) to be assigned a single equivalent thermal conductivity. However, with this narrow extended air space, were its thermal resistance to be based on the larger dimension (typically several metres in length) in the principal direction of heat flow, it would result in

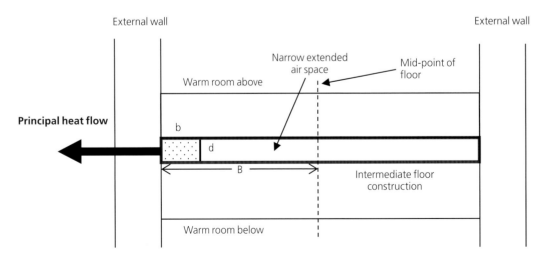

Figure 5: Intermediate floor with a narrow extended air space

a very high (and erroneous) equivalent thermal conductivity. To deal with this, the extended narrow air space is considered to have an equivalent thermal conductivity that is determined from a much smaller length (in the direction of heat flow) that is twice the thickness of the extended air space. For example, in Figure 5 the dimensions of the 'divided' air space are d perpendicular to and b parallel to the principal direction of heat flow and where b = 2 × d. Thus, for the purposes of determining the equivalent thermal conductivity to use for the entire extended air space, this is determined as in section 2.4.2 but for a divided air space of b × d mm.

2.5 Surface heat transfer (surface resistances)

Heat transfer between the internal or external environment and the surfaces of the building is a complex process, depending on a combination of radiation from surrounding materials and convection from air movement over the surface. For practical calculations to be possible, a series of standard coefficients has been developed, which at the inside surface (and sheltered external surfaces) depends on the direction of heat flow. These values, R_{si} and R_{se}, which are quoted as surface thermal resistances in BS EN ISO 6946[5], are shown in Table 1 together with their reciprocals, h_{si} and h_{se}, the surface heat transfer coefficients, either of which are required as inputs to software. The appropriate values from Table 1 should be used in calculations. However, for non-standard surface resistances, such as low emissivity surfaces or external sheltered surfaces, the standard values are replaced with values calculated using the equations in Annex A of BS EN ISO 6946. These calculated non-standard surface resistances (or surface transfer coefficients) should be entered in the model to three significant figures.

Where two different boundary conditions meet, for example at the junction between a wall and a ceiling (Figure 6), each boundary condition should continue as far as possible into the junction of the two boundaries.

In some modelling software the boundary conditions are assigned by areas (2D models) or by volumes (3D models) and, since the boundary conditions cannot overlap (individual nodes can have only one boundary condition), one of the boundary conditions should have a thickness of no more than 1 mm. However, if where the boundaries meet there is also severe thermal bridging, for example metal sections bridging

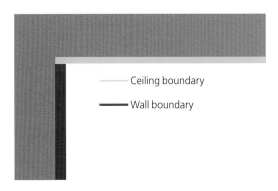

— Ceiling boundary

— Wall boundary

Figure 6: Wall and ceiling boundary conditions in a corner

an insulation layer, then the thickness of the thinner boundary condition should be no more than 0.1 mm.

2.6 Refining the mesh size

When the model of a structure is developed it is usual to specify a 'minimum mesh', ie only the points necessary to define the size and shape of the materials present. Both the heat flows and surface temperatures derived from this will be significantly different from reality. As the model is refined by subdividing the grid, the heat flow and inside surface temperature will both move towards an asymptotic value, as illustrated in Figure 7. It is important therefore to subdivide the grid in the most appropriate manner. Since the level of subdivision of the grid can make a significant difference to the results it is important to use a well-defined and reproducible method.

The easiest way to divide the grid is to simply divide each element into two successively until the calculated heat flow and temperature stabilises. This is straightforward for a relatively simple 2D model, but rapidly becomes unmanageable for a complex 3D structure, where each successive division increases the number of grid points by a factor of eight. It is more important to concentrate the grid elements around the structural elements that are likely to cause significant heat flow.

Table 1: Standard surface resistances and heat transfer coefficients

	Direction of heat flow		
	Upwards	Horizontal*	Downwards
Inside surface			
R_{si} (m²K/W)	0.10	0.13	0.17
h_{si} (W/m²K)	10.0	7.69	5.88
Outside surface			
R_{se} (m²K/W)	0.04	0.04	0.04
h_{se} (W/m²K)	25.0	25.0	25.0

* The values under 'horizontal' apply to heat flow directions ±30° from the horizontal plane (eg if a roof slope is greater than 60°, the horizontal values should be used, otherwise the upwards (or downwards) values are used).

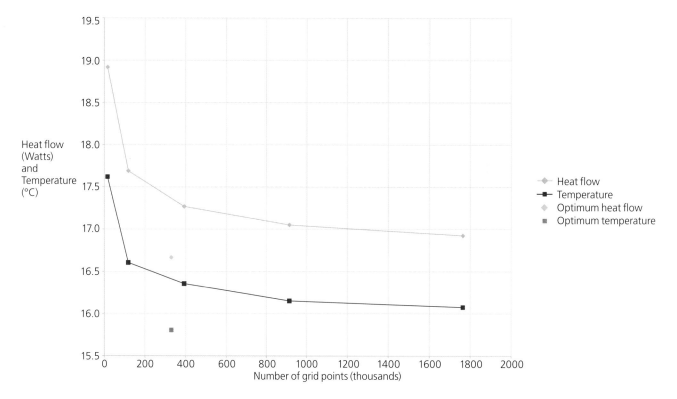

Figure 7: Calculated heat flow and minimum internal temperature in a corner model which is successively subdivided

Figure 7 shows the heat flow and surface temperature through the corner in a hybrid steel-framed wall in which the steel studs penetrate part of the insulation layer.

As the grid is successively subdivided from the minimum grid, the calculated values each fall by about 10%. If, however, the grid is subdivided just once and then extra nodes are inserted around the steel studs to give an 'optimum' subdivision, a significantly lower result occurs[5].

To calculate a sufficiently accurate and reliable thermal model the following procedure is recommended.

1   Define the minimum grid necessary to specify the materials present.

2   Divide all the spaces between the grid points into two.

3   Identify all areas where higher thermal conductivity materials[6] bridge those with lower thermal conductivity, together with areas close to the thermal bridge junction itself, and consider adding more grid points at these locations.

4   Calculate the resulting heat flow and lowest internal surface temperature.

5   Divide all the grid elements into two[7] and recalculate the results.

If the total heat flow calculated from steps **4** and **5** differs by less than 1% and the minimum temperature factor by less than 0.005, the calculation is complete. If the change in either is greater than this, repeat steps **3** to **5**, until the foregoing criterion is met. If the particular modelling software used does not allow the foregoing procedure to be followed (eg if the software has automatic meshing and within that no manual intervention or setting is possible), then the adequacy of the software to produce a sufficiently refined mesh should be demonstrated separately. One way of doing so would be to produce an equivalent of the iteration function of steps **4** and **5** such that subsequent models with different mesh densities can still be manually compared and the stopping criteria of steps **4** and **5** still applied.

2.7 Reporting of temperatures and heat flows

Different numerical modelling programs can report temperatures and heat flows to varying levels of detail. For the determination of temperature factors the minimum requirement is to have all relevant temperatures (external, internal and minimum surface) reported to three decimal places or three significant figures, whichever provides the greater precision. Most programs will graphically display isotherms on the internal surface enabling the location(s) of the minimum temperature factor to be readily determined. For the determination of the linear thermal transmittance, the total heat flow through the model should be reported to four decimal places or four significant figures, whichever provides the greatest precision, and this value should be used in the subsequent calculation of Ψ.

---

[5]   With some modelling packages the asymptote is approached from lower values of heat flow and temperature.

[6]   This includes air spaces, which in the model are assigned an equivalent thermal conductivity.

[7]   For some models, particularly 3D models that include large ground dimensions, the model at step 5 may become too large to compute. If this is the case, the dividing of grid elements at step 5 can be restricted to the grid elements identified at step 3.

3 Thermal bridging at junctions

The severity of a thermal bridge at a junction is indicated by an increased heat loss and reduction in inside surface temperature at the thermal bridge. The heat flow assigned to a linear thermal bridge is represented by its linear thermal transmittance (Ψ-value) in W/m·K and the temperature by the temperature factor, f.

3.1 Ψ-value

3.1.1 Definition

The Ψ-value represents the extra heat flow through the linear thermal bridge over and above that through the adjoining plane elements. From the numerical modelling of a 2D junction, L^{2D}, is the thermal coupling coefficient between the internal and external environments and is calculated from:

$$L^{2D} = \frac{Q}{T_i - T_e} \quad (W/m \cdot K) \tag{1}$$

where:

Q = total heat flow in W/m from the internal to the external environment

T_i = temperature of the internal environment

T_e = temperature of the external environment.

Hence the linear thermal transmittance, Ψ, of the 2D junction is the residual heat flow from the internal to external environment after subtracting the 1D heat flow through all flanking elements, expressed in W/m·K and is determined from:

$$\Psi = L^{2D} - \sum (U \times \ell) \quad (W/m \cdot K) \tag{2}$$

where:

L^{2D} = thermal coupling coefficient

U = U-value (W/m·K) of the flanking element

ℓ = length (m) over which U applies.

Figure 8 shows a 2D horizontal cross-section of the normal corner junction of a simple masonry wall with flanking elements A and B, whence the Ψ-value is calculated from:

$$\Psi = L^{2D} - (U_A \times \ell_A) - (U_B \times \ell_B) \quad (W/m \cdot K) \tag{3}$$

where:

L^{2D} = thermal coupling coefficient

U_A and U_B = U-values (W/m·K) of the flanking elements A and B

ℓ_A and ℓ_B = lengths (m) over which U_A and U_B apply.

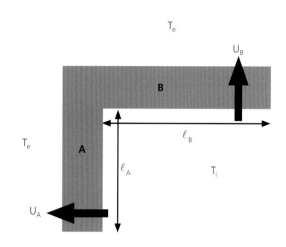

Figure 8: 2D model of a wall corner junction

Note: areas where two or more linear thermal bridges meet, for example at the junction of two walls and a floor, form point thermal bridges, where the heat loss is higher. However, since the area affected is usually small, they are ignored in heat loss calculations.

The heat transmission coefficient, H, of a building is obtained from the sum of the heat flows through *all* of the plane building elements plus that through *all* of the thermal bridges:

$$H = \sum (A \times U) + \sum (\ell \times \Psi) \quad (W/K) \tag{4}$$

3.1.2 Areas over which U-values are applied

For H to be correct, it is important that, when calculating the areas of the plane elements, the dimensions used are consistent with the dimensions used for the corresponding flanking elements in the thermal models. To be consistent, the areas over which the various U-values are applied are the internal finished dimensions of the exposed elements that form the building envelope. For a 2D model of a junction this becomes the length over which the U-value of the particular flanking element applies and for most junction types this is straightforward and unambiguous. However, for some junction types, the extent of this internal area (or length in a 2D model) is not always obvious.

A helpful definition to use for such internal dimension limits is that they should be measured between the finished internal faces of the externally exposed element. In other words, up to the point internally where the exposed element ends externally. This is illustrated with the horizontally staggered party-wall junction (see secton 4.4.3). The only exception to this is where an inverted junction detail is visibly delimited by the internal surfaces of the exposed element(s) turning through an obvious visible angle, for example in the case of the inverted corner junction (see section 4.4.2) where the areas are up to the physical and visible corner.

3.1.3 Modelling U-value, U′

As shown in equation (2), the U-value of the plane (flanking) elements adjacent to the thermal bridge is needed for the calculation of the Ψ-value. Because this calculation involves the difference between two similar (and relatively large) total heat flows, it is sensitive to the precise U-value used in the flanking element. To emphasise this, the concept of the 'modelling U-value', U′, has been introduced. This is the U-value that gives the 1D heat flow that should be subtracted from the total modelled heat flow for each of the flanking elements, and is not necessarily the U-value of the plane building elements that is used for determining the heat loss through them in the context of calculating the total heat loss from the building.

3.1.3.1   Repeating bridging parallel to a junction
Where a flanking element includes repeating thermal bridging that is *parallel* to the thermal bridge junction, if there is a repeating thermal bridge closer to the junction than half of the normal repeat dimension in the flanking element, then that particular, and otherwise repeating, thermal bridge[8] should be included in the modelling of the junction, where its effect on the heat flow is incorporated within the resulting Ψ-value for the junction. All other repeating bridges in the flanking element are *not* included in the model or in the determination of the modelling U-value, U′, of the flanking element.

Figure 9 shows the horizontal cross-section of a steel-frame wall corner junction with vertical steel sections that are *parallel* to the (vertical) corner junction. Thus, in Figure 9 the only repeating steel sections to be included in the model are those in the corner (solid outline), the repeating steel sections away from the corner (dashed outline) are *not* included in the model nor in the determination of the modelling U-value, U′, of the flanking element.

3.1.3.2   Repeating bridging perpendicular to a junction
Where the flanking element has repeating thermal bridging that is perpendicular to the thermal bridge junction, and this repeating thermal bridging significantly affects the heat loss at the thermal bridge junction, it *must* be included in the numerical model. In this case the modelling U-value *must also include* the effect of the repeating thermal bridging, where this is derived from a *separate numerical model*[9] of the plane element that includes the repeating bridging. The effect of repeating thermal bridging (in flanking elements) that is *perpendicular* to the thermal bridge junction can be ignored if:

• the thermal conductivity of the bridging element of the repeating bridge is no greater than 0.5 W/m·K
or
• the bridging element does not bridge the primary insulation layer but bridges instead a lesser thermally resistive layer whose thermal resistance is no greater than 0.2 × the thermal resistance of the primary insulation layer.

Otherwise, the repeating thermal bridging in such a flanking element must be included in any modelling of the junction and in the calculation of the modelling U-value, U′, of the flanking element. For example, the repeating steel bridging elements in a hybrid or cold steel-framed wall cannot be ignored, whereas the steel bridging elements in a warm steel-framed wall can be ignored. Figure 10 shows a wall/floor junction where the vertical steel sections of the cold steel-framed wall are *perpendicular* to the horizontal wall/floor junction.

These steel sections bridge the primary insulation layer and will have a significant effect on the inside surface temperatures and

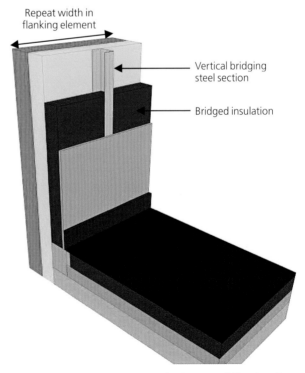

Repeat width in flanking element

Vertical bridging steel section

Bridged insulation

Figure 10: Repeating bridging perpendicular to a wall/floor junction

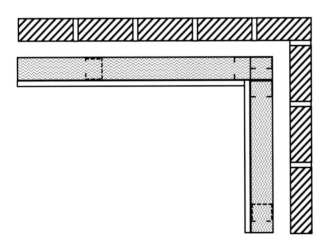

Figure 9: Repeating bridging parallel to wall corner junction

---

8   There may be more than one bridging element within the normal half repeat distance.

9   This separate model can usually be readily constructed by moving the boundaries of the original model to include only the repeating thermal bridge (within the flanking element of interest), ie excluding the thermal bridge junction from the original model.

heat loss of the wall/floor junction itself. They must therefore be included in the modelling of the wall/floor junction and in the calculation of the modelling U-value, U', of the particular flanking element that contains the repeating steel sections. Consequently, the otherwise 2D model of the wall/floor junction must now be modelled in 3D, ie a 3D slice of a 2D junction, where the width of the 3D slice is the repeat of the steel spacing and where the vertical steel in the wall is placed centrally in the width of the slice. The calculated Ψ-value is then the average Ψ-value along the length of the wall/floor junction *between the repeats[10] of the steel sections in the flanking element.*

If the other flanking element also has repeating thermal bridging that cannot be ignored, then the 3D slice of the 2D junction should include the repeating thermal bridging in both flanking elements. Figure 11 shows such a situation.

The width of the 3D slice is then the larger of the repeat lengths from each flanking element. The thermal bridging element in the flanking element with the larger repeat distance should be placed centrally along the width of the 3D slice. In the flanking element that contains the more frequent thermal bridging elements, the 3D slice of the 2D junction may have more than one occurrence of the bridging element in that flanking element, in which case one of these bridging elements should be placed centrally along the width of the 3D slice, so as to 'connect' with the single bridging element in the other flanking element. The remaining bridging element(s) are then located relative to the first at the repeat distance for that flanking element. The one exception to this approach is where the spacing in each flanking element of the repeat of the bridging material(s) is the same, but where one set of bridging elements is intentionally offset from

the other, in which case the offset must be included in the 3D model of the otherwise 2D junction.

3.1.3.3 Mortar joints and other similar repeating bridging within an otherwise homogeneous layer

Repeating thermal bridging contained within a layer of a plane flanking element, such as mortar joints in lightweight blockwork or other similar regular features within an otherwise homogeneous layer, is treated differently from that in section 3.1.3.1 or 3.1.3.2. Such thermal bridging cannot be described either as uniquely parallel or perpendicular to the junction and is often much more frequent (in relation to the thickness of the layer being bridged) compared with the repeating thermal bridging described in section 3.1.3.1 or 3.1.3.2. The thermal bridging effects are instead taken into account (in the model) by assigning an equivalent thermal conductivity to the blockwork (or other inhomogeneous layer). The equivalent thermal conductivity is obtained from dividing the thickness of the layer (m) by the thermal resistance of the layer (m²K/W) calculated using the combined method of BS EN ISO 6946[5] and using a surface resistance of zero on both sides of the layer. *Note:* for the common situation of mortar with dense blockwork or brick, mortar joints can be ignored (see section 4.2 of BRE Report BR 443[10]).

3.1.4 Alternative expression for calculating Ψ

An alternative expression for the definition and calculation of Ψ is useful when ΔT across one or more flanking elements in a particular modelled junction is not the full temperature difference between the internal and external environments (ie ΔT does not equal $(T_i - T_e)$). This occurs for models of junctions that have a third temperature boundary in the model that is different from the internal or external temperature. This alternative expression for the calculation of Ψ is as follows.

For a 2D model, the equation is:

$$\Psi = \frac{Q^{2D} - \sum (U \times \Delta T \times \ell)}{(T_i - T_e)} \text{ (W/m·K)} \tag{5}$$

where:

Q^{2D} = total heat flow through the 2D model (W/m)

U = U-value of the flanking element (W/m·K)

ΔT = temperature difference (°K) across the flanking element

ℓ = length in m over which the U-value applies

T_i = internal temperature (°C)

T_e = external temperature (°C).

For a 3D model, equation (5) becomes:

$$\Psi = \frac{Q^{3D} - \sum (U \times \Delta T \times \ell \times w)}{(T_i - T_e) \times w} \text{ (W/m·K)} \tag{6}$$

where:

Q^{3D} = total heat flow through the 3D model (W)

U = U-value of the flanking element (W/m²K)

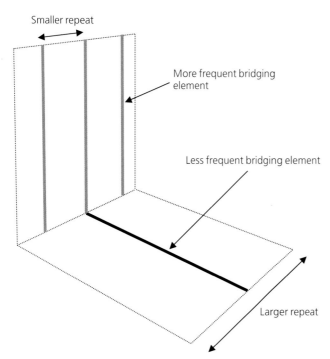

Smaller repeat

More frequent bridging element

Less frequent bridging element

Larger repeat

Figure 11: Repeating bridging in both flanking elements of a junction

---

[10] In general, most walls will have several of these repeats along the length of the junction with the resulting Ψ-value applied to the whole length of the junction, irrespective of there being an integer number of repeats along it.

ΔT = temperature difference (°K) across the flanking element

ℓ = length (m) over which the U-value applies

T_i = internal temperature (°C)

T_e = external temperature (°C)

w = width of the model (m).

3.2 Temperature factor, f_{Rsi}

The temperature factor, f_{Rsi}, is used to assess the risk of surface condensation or mould growth on any detail. It is calculated (under steady-state conditions) from:

$$f_{Rsi} = \frac{T_{si} - T_e}{T_i - T_e} \qquad (7)$$

where:

T_{si} = surface temperature

T_i = internal environmental temperature

T_e = temperature of the external environment.

f_{Rsi} depends only on the construction and not on the imposed air temperatures. Once it has been calculated for any particular T_i and T_e, it can be used to calculate the surface temperature for any other set of conditions using:

$$T_{si} = T_e + f_{Rsi}(T_i - T_e) \qquad (8)$$

The value of internal surface resistance (R_{si}) assumed has a very significant effect on the calculated surface temperature. It is therefore essential that the surface resistance appropriate to the element being analysed (see section 2.5) is used at all times.

To avoid problems of surface condensation or mould growth f should be not less than the critical temperature factor f_c. BRE Information Paper IP 1/06[7] gives appropriate values for f_c for different internal environments[11].

3.2.1 Temperature factors for ground-floor junctions

There are generally two temperature factors that are relevant for junctions involving ground floors. The first is that obtained from a 2D model of the wall/ground-floor junction. If the adjacent wall/ground-floor junction is different, then a second 2D model will be required to provide the relevant temperature factor for that 2D model. If there is variation in the cross-section of either of the wall/ground-floor junctions (this may arise where there are repeating thermal bridges in the wall and/or floor), then a

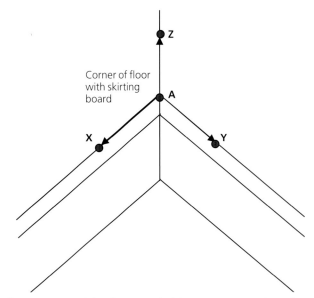

Corner of floor with skirting board

Figure 12: Determining the reported minimum temperature factor for a ground-floor corner. Point A, is, for example, the coldest point. The other points, X, Y and Z, are 10 mm away from point A

3D model of the wall/ground-floor junction will be required, for determining both the 'average' Ψ-value and the minimum temperature factor for that junction.

The second temperature factor is obtained from a 3D model of the wall/wall/ground-floor junction. Often, there is a severe dip in internal surface temperature in the corner of a ground floor where the adjacent walls and floor meet. Where skirting boards are present, which is usually the case, the minimum surface temperature is not always at the corner of the wall and floor, but can be transferred to the corner of the two walls immediately above the skirting.

When assessing the minimum surface temperatures in these two locations, for regulatory purposes there is a relaxation in distance from these two corner junctions of 10 mm (in all directions) within which the surface temperatures need not be considered, when determining points with the minimum temperature factor. In other words, only temperature factors beyond these 10 mm zones need be considered when comparing these to the critical surface temperature factor as given in BRE IP 1/06[7].

The reported corner temperature factor is obtained from the minimum of the three temperatures that are 10 mm away (in each of the three orthogonal axes, ie X, Y, and Z in the model) from the coldest temperature in the corner, point A (Figure 12).

---

[11] *Note:* the values of f_c in BRE IP 1/06[7] have been set not only to take account of the appropriate humidity levels of the different classes of building, but also to compensate for the use of lower internal surface resistances when modelling to the guidance given here, compared with modelling at the higher value of 0.25 m²K/W (for all internal surfaces) recommended for corners (see 4.4.1 of BS EN ISO 13788:2012[12]).

4 Junction types

The various junctions between plane building elements can be grouped as listed below and in the relevant sections that follow.

- Roof junctions (section 4.1)
- Room-in-roof junctions (section 4.2)
- Junctions around openings in the external wall or roof (section 4.3)
- Corner junctions (section 4.4)
- Intermediate-floor or party-wall junctions (with the external wall) (section 4.5)
- Exposed floors (section 4.6)
- Ground-floor junctions (section 4.7)
- Special-case junctions that connect to the ground (section 4.8)

Some of the modelling conventions that follow have similarities across most of the above groups. Others relate to the specific group and still others are peculiar to specific junction types within each group. Listed below are the types of junction with the conventions that apply, together with the equation(s) to use when determining Ψ from the output of the numerical model(s).

In the drawings that follow, the red line depicts the usual warm internal boundary (usually set to 20 °C), the blue line is the external boundary (usually set to 0 °C) and the green line is an adiabatic (zero heat flow) boundary.

Where there are other environments, for example, well ventilated soffits at wall/roof junctions, loft spaces or underfloor spaces, other colours denote the different boundary conditions that apply.

Certain of the junction types discussed below can involve one or more properties. Where this is the case the Ψ-value is determined for and applies to the junction as a whole and it should be made clear by the modeller (for subsequent users of the particular calculated Ψ-value) that this is the case. The proportion of the Ψ-value to apply to the different properties involved in a particular build is decided on separately in accordance with the conventions that are being applied when calculating the total fabric heat loss for a property. Often such a junction is shared typically between only two properties, in which case half of the Ψ-value determined for the whole junction would be applied to each property.

Where a junction heat loss is to an unheated space, such as an integral garage or a stairwell, the unheated space is modelled as though it was fully exposed to the external environment. Subsequent use of the reported Ψ-value from this modelling would have the Ψ-value reduced in accordance with the conventions that are being applied when calculating the total fabric heat loss for a property.

4.1 Roof junctions

4.1.1 Roof eaves (insulated at ceiling)

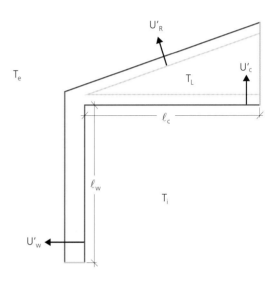

$$\Psi = \frac{Q - U'_W \times \ell_W \times (T_i - T_e) - U'_C \times \ell_C \times (T_i - T_L)}{(T_i - T_e)}$$

The temperature of the loft space, T_L, is considered to be equal to T_e, the temperature of the external environment (see section 2.2.5.2). U'_C is the U-value between the room and the loft space, where R_{se} of the upper surface is taken to be 0.10 m²K/W. The sloping roof construction[12] is not included in the model, but the boundary conditions defined by its presence are included. Similarly, where the eaves are ventilated (which is usually the case), the air space behind the soffit at the eaves is considered to be still external air, giving a boundary condition for the surfaces behind the soffit of T_e (°C) and with a surface resistance equal to that for still air.

See *Worked example 1* in Appendix B.

---

[12] An exception is where the rafters are made of metal, in which case the sloping part of the roof must be included in the model. *See also* the guidance in section 3.1.3 about repeating thermal bridging in flanking elements.

4.1.2 Roof gable (insulated at ceiling)

4.1.3 Roof/party wall (insulated at ceiling)

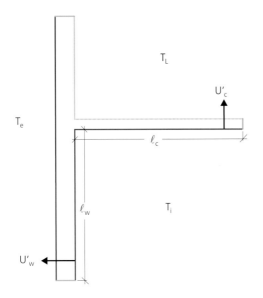

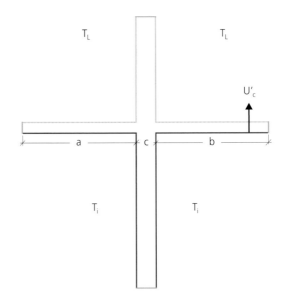

$$\Psi = \frac{Q - U'_w \times \ell_w \times (T_i - T_e) - U'_c \times \ell_c \times (T_i - T_L)}{(T_i - T_e)}$$

$$\Psi = \frac{Q - U'_c \times \ell_c \times (T_i - T_L)}{(T_i - T_L)}$$

As for secton 4.1.1, the temperature of the loft space ,T_L, is considered to be equal to T_e, the temperature of the external environment (see section 2.2.5.2) and U'_c is again calculated with an R_{se} of 0.10 m²K/W.

Note: the required minimum length in the model for the gable wall to the loft space is determined following the rules in section 2.2.2.

As for section 4.1.1, the temperature of the loft space, T_L, is considered to be equal to T_e, the temperature of the external environment (see section 2.2.5.2) and U'_c is again calculated with an R_{se} of 0.10 m²K/W.

For a separating wall in the same dwelling (or premises), the length ℓ_c over which the U-value U'_c applies is:

$$\ell_c = a + b + c$$

For a party wall between different dwellings (or premises), the length ℓ_c over which the U-value U'_c applies is:

$$\ell_c = a + b$$ with half of Ψ allocated to each dwelling (or premises).

Note: the required minimum length in the model for the party wall (above and below) the ceiling is determined following the rules in section 2.2.2.

4.1.4 Roof eaves (insulated on slope)

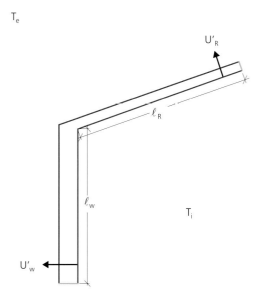

$$\Psi = \frac{Q - U'_W \times \ell_W \times \left(T_i - T_e\right) - U'_R \times \ell_R \times \left(T_i - T_e\right)}{\left(T_i - T_e\right)}$$

4.1.5 Roof gable (insulated on slope)

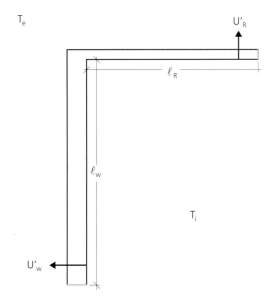

$$\Psi = \frac{Q - U'_W \times \ell_W \times \left(T_i - T_e\right) - U'_R \times \ell_R \times \left(T_i - T_e\right)}{\left(T_i - T_e\right)}$$

For a sloping roof, in the 2D model, a strictly 2D cross-section through the party wall and its connection with the sloping roof would produce an angled cross-section through the sloping roof construction and would result in greater (and hence incorrect) thicknesses through the sloping roof construction. A fully correct model would require the junction to be modelled using a 3D model. However, a 2D model is considered adequate, provided the thicknesses through the roof construction are the actual thicknesses. For the purposes of the modelling, this is equivalent to the sloping roof construction being considered to be a flat roof. If the roof is a flat roof, then the actual construction thicknesses of the flat roof construction apply (correctly) to the 2D cross-section of the 2D model.

4.1.6 Roof/party wall (insulated at roof level)

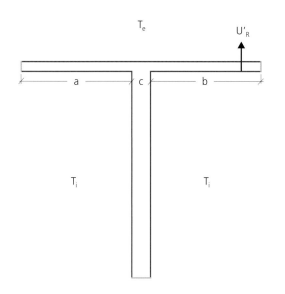

$$\Psi = \frac{Q - U'_R \times \ell_C \times (T_i - T_e)}{(T_i - T_e)}$$

As for section 4.1.5, a 2D model is considered adequate to describe the connection of the sloping roof with the party wall, provided the thicknesses through the roof construction are the actual thicknesses. For the purposes of the modelling, this is equivalent to the sloping roof construction being considered to be a flat roof. If the roof is a flat roof, then the actual construction thicknesses of the flat roof construction apply (correctly) to the two-dimensional cross-section of the 2D model.

For a separating wall in the same dwelling (or premises), the length ℓ_C over which the U-value U'_R applies is:

$$\ell_C = a + b + c$$

For a party wall between different dwellings (or premises), the length ℓ_C over which the U-value U'_R applies is:

$\ell_C = a + b$, with half of Ψ allocated to each dwelling (or premises).

Note: the required minimum length in the model for the party wall is determined following the rules in section 2.2.2.

4.1.7 Flat-roof eaves

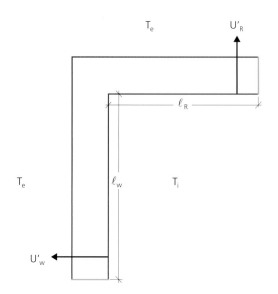

$$\Psi = \frac{Q - U'_W \times \ell_W \times (T_i - T_e) - U'_R \times \ell_R \times (T_i - T_e)}{(T_i - T_e)}$$

4.1.8 Flat-roof parapet

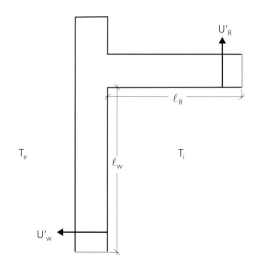

$$\Psi = \frac{Q - U'_W \times \ell_W \times (T_i - T_e) - U'_R \times \ell_R \times (T_i - T_e)}{(T_i - T_e)}$$

4.2 Room-in-roof junctions

The following room-in-roof junctions involve connections between plane elements that make up the room-in-roof. These junctions are formed from various connections of vertical walls with flat ceilings or with flat or sloping roofs, or flat ceilings or roofs with sloping roofs. In the case of spaces to the outside of the insulation envelope, such as side walls to the space between them and a sloping roof or the space between an insulated dormer ceiling and a small loft space, the temperature of these spaces (and hence boundary conditions in the model) is taken to be that of the external environment, as though these spaces were fully exposed to the external air. Similarly, for junction type 4.2.3, where the residual loft space is well insulated at ceiling level, its temperature is considered to be equal to that of the external air, ie equal to T_e (see section 4.1.1). Where a property has a vaulted ceiling (rather than a room-in-roof), certain roof junctions are treated as for room-in-roof. For example, the ridge junction of a vaulted ceiling is treated as for the ridge junction of a room-in-roof (see section 4.2.1).

4.2.1 Ridge (vaulted ceiling)

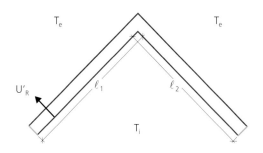

$$\Psi = \frac{Q - U'_R \times (\ell_1 + \ell_2) \times (T_i - T_e)}{(T_i - T_e)}$$

4.2.2 Ridge (vaulted ceiling – inverted)

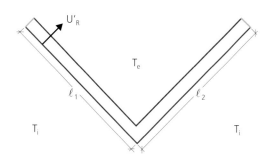

$$\Psi = \frac{Q - U'_R \times (\ell_1 + \ell_2) \times (T_i - T_e)}{(T_i - T_e)}$$

4.2.3 Sloping rafter to flat ceiling (insulated at ceiling)

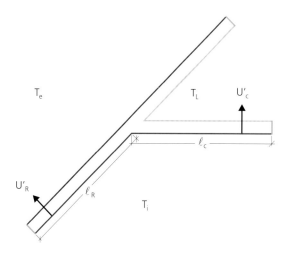

$$\Psi = \frac{Q - U'_R \times \ell_R \times (T_i - T_e) - U'_C \times \ell_C \times (T_i - T_L)}{(T_i - T_e)}$$

The temperature of the loft-space ,T_L, is considered to be equal to T_e, the temperature of the external environment (see section 2.2.5.2).

Note: the sloping roof construction adjacent to the loft space is included in the model where its required minimum length is determined following the rules in section 2.2.2.

See *Worked example 2* in Appendix B.

4.2.4 Flat ceiling to rafter slope (inverted)

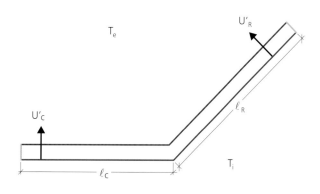

$$\Psi = \frac{Q - U'_C \times \ell_C \times (T_i - T_e) - U'_R \times \ell_R \times (T_i - T_e)}{(T_i - T_e)}$$

4.2.5 Roof wall to rafter

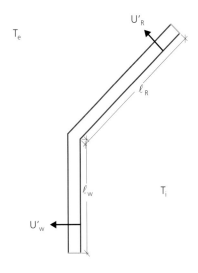

$$\Psi = \frac{Q - U'_W \times \ell_W \times (T_i - T_e) - U'_R \times \ell_R \times (T_i - T_e)}{(T_i - T_e)}$$

4.2.6 Roof wall to flat ceiling

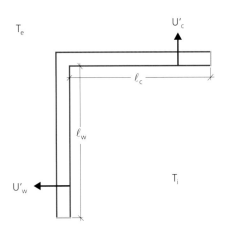

$$\Psi = \frac{Q - U'_W \times \ell_W \times (T_i - T_e) - U'_c \times \ell_c \times (T_i - T_e)}{(T_i - T_e)}$$

4.3 Junctions around openings

These are the junctions around the openings in the walls (or roofs) of buildings that receive the windows and doors as appropriate. In the case of openings in walls the specific junction types are the lintel, sill and jamb of the window and the lintel, threshold[13] and jamb of the door. In the case of a roof window or rooflight the junction detail can be similar in type to the jamb of a window or door. In the case of some rooflights, however, these can also involve upstands. In the various outline illustrations that follow, the frame of the door, window or rooflight is indicated with a black dashed line (to aid understanding) but the frame itself is not part of the model of the junction.

4.3.1 Lintel (window head)

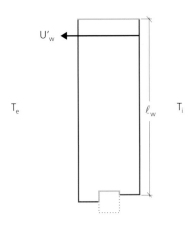

$$\Psi = \frac{Q - U'_W \times \ell_W \times (T_i - T_e)}{(T_i - T_e)}$$

Some modelling software can only cope with rectangular geometries and for metal lintels, where one or more faces of the lintel is sloping, the sloping part is modelled by an arrangement of steps along the length of the slope determined according to section 2.2.4. If there is a perforated plate to the lintel (see section 2.3.1) the equivalent thermal conductivity should be determined by separate numerical modelling and this thermal conductivity used for the 'solid' plate.

See *Worked example 3* in Appendix B.

---

[13] A door threshold that is in connection with the ground is dealt with under section 4.7, *Ground-floor junctions*.

4.3.2 Jamb

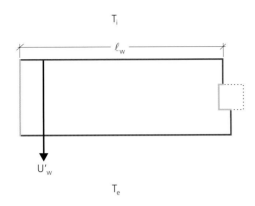

$$\Psi = \frac{Q - U'_W \times \ell_W \times (T_i - T_e)}{(T_i - T_e)}$$

4.3.3 Sill

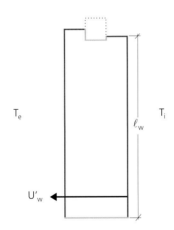

$$\Psi = \frac{Q - U'_W \times \ell_W \times (T_i - T_e)}{(T_i - T_e)}$$

Note: the length ℓ_W includes the thickness of any internal sill.

4.3.4 Rooflight

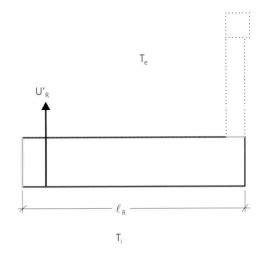

$$\Psi = \frac{Q - U'_R \times \ell_R \times (T_i - T_e)}{(T_i - T_e)}$$

As with other openings, there is an adiabatic boundary in the model where the frame of the rooflight (or the upstand if present) connects with the roof.

4.4 Corner junctions

The following junctions are the vertical junction between two adjacent external walls.

4.4.1 Corner (normal)

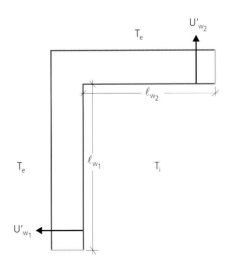

$$\Psi = \frac{Q - U'_{W_1} \times \ell_{W_1} \times (T_i - T_e) - U'_{W_2} \times \ell_{W_2} \times (T_i - T_e)}{(T_i - T_e)}$$

4.4.2 Corner (inverted)

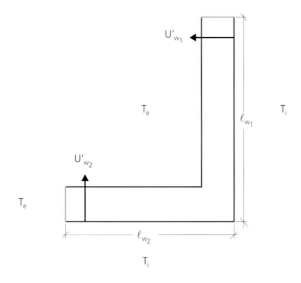

$$\Psi = \frac{Q - U'_{W_1} \times \ell_{W_1} \times (T_i - T_e) - U'_{W_2} \times \ell_{W_2} \times (T_i - T_e)}{(T_i - T_e)}$$

Note: if there are no constructional aspects to the thermal bridging at a corner junction (which is usually the case when the insulation layer is continuous around the inverted corner) the Ψ-value for the inverted corner will have a lower Ψ-value than that for the corresponding normal corner, and indeed it can be negative.

4.4.3 Staggered party wall (horizontal)

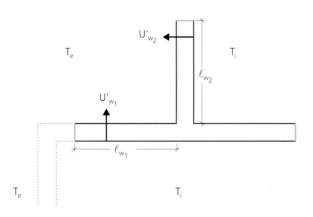

$$\Psi = \frac{Q - U'_{W_1} \times \ell_{W_1} \times (T_i - T_e) - U'_{W_2} \times \ell_{W_2} \times (T_i - T_e)}{(T_i - T_e)}$$

Note: the required minimum length in the model for the party wall is determined following the rules in section 2.2.2.

See *Worked example 4* in Appendix B.

4.5 Intermediate-floor or party-wall junctions (with external wall)

4.5.1 Intermediate floor

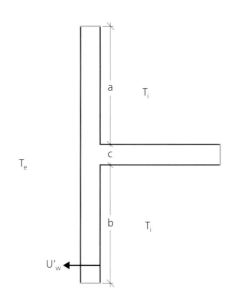

$$\Psi = \frac{Q - U'_W \times \ell_W \times (T_i - T_e)}{(T_i - T_e)}$$

For an intermediate floor in the same dwelling (or premises), the length ℓ_W over which the U-value U'_W applies is:

$\ell_W = a + b + c$

For an intermediate floor in different dwellings (or premises):

$\ell_W = a + b$ with half of Ψ allocated to each dwelling (or premises).

Note: the required minimum length in the model for the intermediate floor is determined following the rules in section 2.2.2.

4.5.2 Balcony

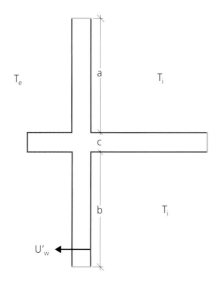

$$\Psi = \frac{Q - U'_W \times \ell_W \times (T_i - T_e)}{(T_i - T_e)}$$

For a balcony and intermediate floor within the same dwelling (or premises), the length ℓ_W over which the U-value U'_W applies is:

$\ell_W = a + b + c$

For a balcony and intermediate floor in different dwellings (or premises), the length ℓ_W over which the U-value U'_W applies is:

$\ell_W = a + b$ with half of Ψ allocated to each dwelling (or premises).

Note: the required minimum length in the model for the intermediate floor is determined following the rules in section 2.2.2.

See *Worked example 5* in Appendix B.

4.5.3 Partition/party wall

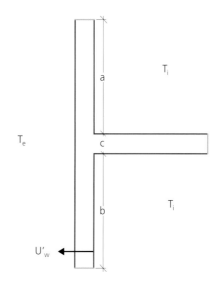

$$\Psi = \frac{Q - U'_W \times \ell_W \times (T_i - T_e)}{(T_i - T_e)}$$

For internal partitions within the same dwelling (or premises), the length ℓ_W over which the U-value U'_W applies is:

$\ell_W = a + b + c$

For a party wall between dwellings (or premises), the length ℓ_W over which the U-value U'_W applies is:

$\ell_W = a + b$ with half of Ψ allocated to each dwelling (or premises).

Note: the required minimum length in the model for the partition/party wall is determined following the rules in section 2.2.2.

4.6 Exposed floors

4.6.1 Exposed floor (normal)

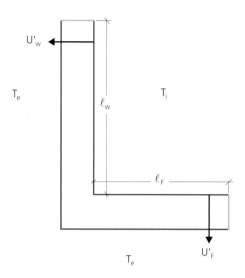

$$\Psi = \frac{Q - U'_w \times \ell_w \times (T_i - T_e) - U'_F \times \ell_F \times (T_i - T_e)}{(T_i - T_e)}$$

4.6.2 Exposed floor (inverted)

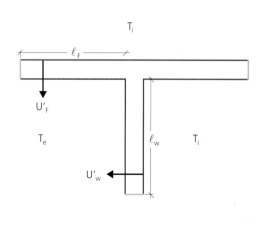

$$\Psi = \frac{Q - U'_w \times \ell_w \times (T_i - T_e) - U'_F \times \ell_F \times (T_i - T_e)}{(T_i - T_e)}$$

Note: the required minimum length in the model for the intermediate floor is determined following the rules in section 2.2.2.

See *Worked example 6* in Appendix B.

4.7 Ground-floor junctions

The following junctions show external wall connections with a ground floor. Clearly, the perimeter of the ground floor will usually include opening thresholds. At threshold junctions the height of any residual wall construction will be much less than three times its thickness. Indeed ℓ_w will usually be zero and any 'step' or 'residual wall' to the threshold detail is included in the model with an appropriate adiabatic where the threshold connects with the (closed) door, ie similar to that of a sill meeting with the window frame (see section 4.3.3). For threshold details, the formulae given for the calculation of Ψ in this section are therefore modified with the second term $U'_w \times \ell_w \times (T_i - T_e)$ deleted. *Note:* any floor skirting should be included in the following models.

For the purposes of determining the temperature factor for the corner of a ground floor (see section 3.2.1), the required 3D model of the floor will not usually involve any repeating thermal bridging elements that are present. However, if there is a repeating bridging element(s) in the floor that is sufficiently close to the corner, this may require to be included in the model (see section 3.1.3).

In section 4.7.1 when calculating U'_f of the ground floor and in section 4.7.2 when calculating T_u (the heat balance temperature of the underfloor space), if the floor construction has horizontal all-over insulation then any edge insulation (vertical or horizontal) is ignored in these calculations of U'_f and T_u. If the floor is insulated using edge insulation alone, the edge insulation is included in the ground-floor calculations. *Note:* in *both* cases, any edge insulation present is still included in the various models.

4.7.1 Solid ground floor

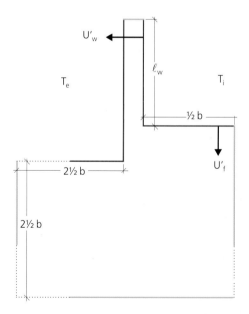

$$\Psi = \frac{Q - U'_w \times \ell_w \times (T_i - T_e) - U'_f \times \tfrac{1}{2} b \times (T_i - T_e)}{(T_i - T_e)}$$

U'_f is calculated in accordance with BS EN ISO 13370[8]. This procedure is implemented in the BRE U-value calculator[9]. The dimension b is, from BS EN ISO 13370, the characteristic dimension of the floor. b in the model should be set at 8 m, which corresponds to a P/A of 0.25 (see section 2.2.2), and the level of the external soil (in the model) should be set 150 mm below that of the internal floor finish.

4.7.2 Suspended ground floor

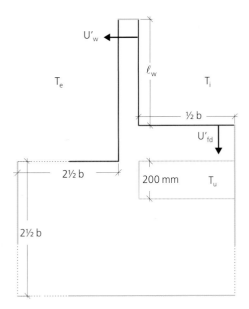

$$\Psi = \frac{Q - U'_w \times \ell_w \times \left(T_i - T_e\right) - U'_{fd} \times \frac{1}{2}b \times \left(T_i - T_u\right)}{\left(T_i - T_e\right)}$$

The underfloor space is at an intermediate temperature, T_u, between T_i and T_e. T_u should be calculated from the heat balance in accordance with Annex E of BS EN ISO 13370[8] *for a P/A of 0.5*[14]. U'_{fd} is the U-value of the floor deck. If the floor deck has *no repeating thermal bridging*, U'_{fd} can be calculated using BS EN ISO 6946[5], otherwise it *must* be determined from a separate model of the floor deck. The dimension b, is (from BS EN ISO 13370) the characteristic dimension of the floor. Dimension b in the model should be set at 8 m, which corresponds to a P/A of 0.25 (see section 2.2.2). The level of the external soil (in the model) should be set 150 mm below that of the internal floor finish and this dimension is also used in the calculations carried out in accordance with Annex E of BS EN ISO 13370. T_u (above) can be calculated using the BRE

U-value calculator. *Note:* when calculating T_u, the ventilation of the underfloor space should be determined in accordance with Annex E of BS EN ISO 13370 assuming a wind speed of 5 m/s, a shielding factor of 0.05 and ventilation openings equivalent to 0.0015 m² per metre length of the floor perimeter.

If the floor deck contains thermal bridging, but a particular model for a floor edge does not require the repeating thermal bridging element of the floor deck to be part of the model, the underfloor temperature T_u (in that model) is determined as above (for the case of the repeating bridging being present), but note that U'_{fd} for this particular model is the U-value of the floor deck *without any repeating thermal bridging elements present* in this (P/A = 0.25) model. Also, for the purposes of determining the temperature factor for the corner of such a suspended floor, the 3D model required for this should also assign T_u (determined as above) to the temperature of the underfloor space.

See *Worked example 7* in Appendix B.

4.7.3 Basement wall/floor junction

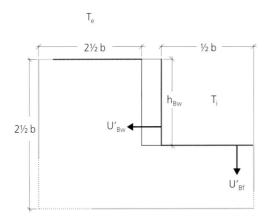

$$\Psi = \frac{Q - U'_{Bw} \times h_{Bw} \times \left(T_i - T_e\right) - U'_{Bf} \times \frac{1}{2}b \times \left(T_i - T_e\right)}{\left(T_i - T_e\right)}$$

U'_{Bw} and U'_{Bf} are calculated in accordance with BS EN ISO 13370. The dimension b, from BS EN ISO 13370, is the characteristic dimension of the floor. Dimension b in the model should be set at 8 m (see section 2.2.2) and h_{Bw} should be set at 2.4 m.

Note: there is no requirement to determine a temperature factor at the 'corner' of a basement floor.

See *Worked example 8* in Appendix B.

---

14 This is representative of a typical size of floor for a dwelling and provides an appropriate underfloor temperature to use subsequently in *all* models that are used to determine the thermal performance of the junction, ie the determination of Ψ and *f*. *Note:* the underfloor temperature for larger floors will be higher compared to that for small floors and the use of P/A = 0.5 here will yield slightly pessimistic results.

4.8 Special-case junctions that connect to the ground

The following junctions have a complex interaction of heat flows to and through the connecting ground and the Ψ-value cannot be adequately determined from a single model (2D or 3D), but instead requires modelling results from two 3D models. These 3D models include the large ground volumes that are required in order to take proper account of the heat flow through these junctions.

4.8.1 Party wall/ground-floor junction

Party walls or separating walls can separate different built forms: mid-terrace, end-terrace and semi-detached. However, the Ψ-value is largely independent of built form with a variation of about + 0.01 W/m·K between the mid-terrace and the semi-detached cases. By convention, the build form to use when modelling the ground-floor/separating wall junction is that of the end-terrace and the method here of calculating the Ψ-value

of the separating wall junction with the ground floor (for the case of the end-terrace) is applied to all built forms that contain this junction type.

To determine the Ψ-value for the junction of a party wall with the ground floor it is necessary to construct two models:

- first, the floor with the separating wall (Figure 4.8.1a), giving the modelled heat loss, Q_1^{3D}
- secondly the floor without the separating wall (Figure 4.8.1b), giving the modelled heat loss, Q_2^{3D}.

Note: both models are identical (ie have the same overall dimensions (areas and volumes)), except that the second model has the party wall and its foundation in the ground removed and replaced by a continuation of the floor construction.

To avoid the Ψ-value of the separating wall junction with the ground floor being dependent on the construction of the external walls, in the models the external walls are replaced by an adiabatic boundary at the edge of the floor (as shown in Figure 4.8.1c) which is vertically down from the finished floor

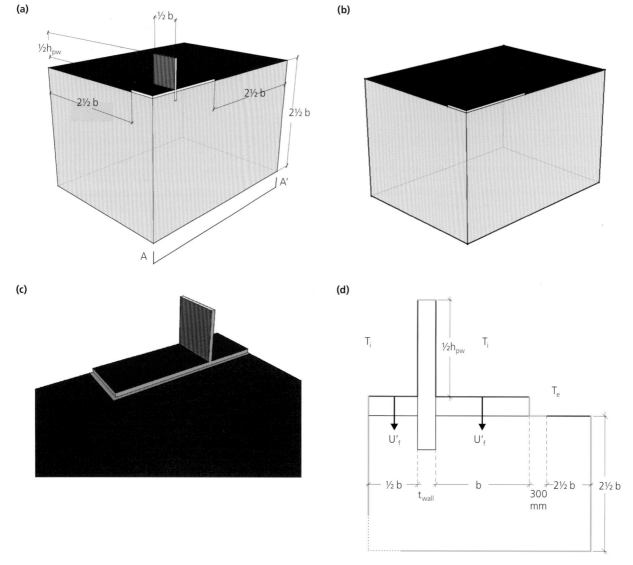

Figure 4.8.1: (a) Complete model of ground floor and separating wall (showing ground dimensions), (b) same model as in (a), but with the separating wall (and its foundation) removed and replaced by the floor construction, (c) model as in (a), but viewed from the opposite viewpoint, (d) 2D cross-section A–A′ from (a)

level and then 300 mm horizontally at the level of the external soil, which is itself 150 mm below the finished floor level.

The cross-section[15] of the end-terrace built form is shown in Figure 4.8.1d. The characteristic dimension, b, of the floor is taken to be 8 m. The width of the floor in the model parallel to the separating wall from the floor edge to the opposite adiabatic is then half of the characteristic dimension of the floor, ie ½b = 4 m. The height of the party wall, h_{Pw}, is taken to be 2.4 m. The height from the finished floor level to the adiabatic at the top of the separating wall in the model is ½h_{Pw}, ie 1.2 m.

The Ψ-value for this junction type has a constructional component and a geometrical component (see below). The constructional Ψ-value, Ψ_c, is calculated from:

$$\Psi_c = \frac{\left(Q_1^{3D} - Q_2^{3D}\right)}{(T_i - T_e) \times \ell} \quad \text{W/m·K} \qquad 4.8.1\ (1)$$

where:

Q_1^{3D} = total heat loss from the model that includes the separating wall (W)

Q_2^{3D} = total heat loss from the model with the separating wall removed (W)

ℓ = width of floor (in the model) parallel to the separating wall (m).

As indicated, the Ψ-value component determined from equation 4.8.1 (1), ie from the two models (with and without the separating wall) relates only to the constructional thermal bridging effect of the junction of the separating wall with the floor. Where the separating wall is between properties, since the U-value of the floor is applied up to the finished internal area of the floor (ie only up to the separating wall), there is an additional unaccounted for heat flow through the junction where this is equal to the U-value of the floor in the model multiplied by the thickness of the separating wall. In the case of a separating wall within the same property (eg a partition wall) there is no additional heat flow since the U-value of the floor is applied to an area that includes the width of the separating wall.

Thus, in the case of a separating wall *between* properties, Ψ is determined from:

$$\Psi = \Psi_c + \left(t_{wall} \times U_f'\right) \quad \text{W/m·K} \qquad 4.8.1\ (2)$$

where:

Ψ_c = value calculated from equation 4.8.1 (1)

t_{wall} = thickness of the separating wall and

U_f' = U-value of the heat-loss floor determined from:

$$U_f' = \frac{Q_f}{\left(b + \tfrac{1}{2}\,b + t_{wall}\right) \times \ell \times \left(T_i - T_e\right)} \qquad 4.8.1\ (3)$$

where:

$Q_f = Q_2^{3D}$, the heat flow into the heat-loss floor of the second 3D model.

Half of the calculated Ψ-value is assigned to each property.

In the case of a separating wall *within* a property, Ψ is equal to Ψ_c from equation 4.8.1 (1) and the full calculated Ψ-value is assigned to the property.

See *Worked example 9* in Appendix B.

Junction type 4.8.1 for a suspended floor

Where the ground floor for this junction type is a suspended ground floor (rather than solid), the overall approach is essentially the same as for the solid ground floor, but with the model containing now a defined underfloor space whose depth is 200 mm measured from the underside of the floor-deck construction. The adiabatic edges of the floor are amended to take into account the presence of the underfloor space. In the model the temperature of the underfloor space is determined as for junction detail 4.7.2. The underfloor space and the amended adiabatic boundary at the edges of the floor are shown in the following equivalent Figures 4.8.1e, f and g that correspond to those of 4.8.1a, b and d from the solid floor case.

As for the case of the solid floor, in order to avoid the Ψ-value of the separating wall junction with the suspended ground floor being dependent on the construction of the external walls, in the models the external walls are replaced with an adiabatic boundary at the edge of the floor (as shown in Figure 4.8.1g) vertically down from the finished floor level and then 300 mm horizontally on top of and at the level of the external soil, which itself is 150 mm below the finished floor level. In the model, any remaining (ie non-adiabatic) vertical boundary between the underfloor space and the external soil is given by a boundary condition of temperature T_U and the usual 'horizontal' surface resistance of 0.13 m²K/W.

The Ψ-value for this suspended floor junction is calculated (as for the solid wall case) using equations 4.8.1 (1), (2) and (3) as appropriate for a separating wall between properties (equation 4.8.1 (1)) or a separating wall within the same property (equation 4.8.1 (2)), where:

Q_1^{3D} = total heat flux from the first model, as indicated by Figure 4.8.1e

Q_2^{3D} = total heat flux from the second model (which has the separating wall and its foundation removed and replaced with the suspended floor construction and soil) as indicated by Figure 4.8.1f.

---

[15] If along the length of the junction there is any repeating thermal bridging in any or all of the flanking elements then the repeating thermal bridging element(s) should be included in the model. These should be located with their usual centre-to-centre spacing, beginning with the precise location of the first of the repeating elements at the end of the junction which is towards the adiabatic replacement for the external wall. *Note:* the length of the junction is still fixed at 4 m (= ½b).

(e)

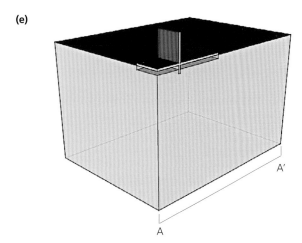

(f)

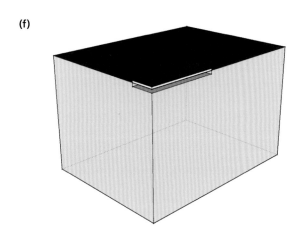

(g)

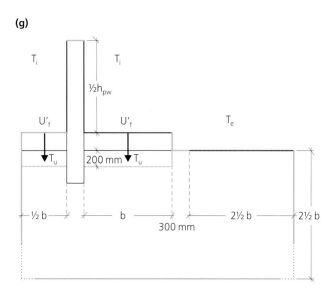

Figure 4.8.1: (e) Complete model of suspended ground floor and separating wall, (f) same model as in (e), but with the separating wall (and its foundation) removed and replaced by the floor construction, (g) 2D cross-section A–A' taken from (e)

4.8.2 Party wall/ground-floor (inverted) junction

This junction is described as an 'inverted' junction, meaning that the internal warm 'area' to the junction is greater than the 'external' cold area, ie the area that connects to the cold ground.

Figure 4.8.2a shows the extent of the ground and its connection to the warm rooms above and below ground. The basement floor is shown in dark brown (marked B_F in the key) and the 'adjacent' basement wall is shown in light brown (marked B_W in the key). *Note:* the basement-floor construction continues under both the basement wall that is part of the junction and also under the 'adjacent' basement wall.

So that the Ψ-value of this junction is independent of the precise construction of both the basement floor and the adjacent basement wall, the real basement floor is replaced in the model by a single layer 'floor' 150 mm thick and the real 'adjacent' basement wall is replaced by a single layer 'wall' that is 300 mm wide and which extends horizontally from the vertical adiabatic of the 'adjacent' wall up to the inside surface of the basement wall of the junction. If the real basement floor is insulated, then the thermal conductivity of the 150 mm thick 'replacement' basement floor in the model is taken to be 0.11 W/m·K, otherwise it is taken to be 1.88 W/m·K. If the real 'adjacent' basement wall is insulated, then the thermal conductivity of the 300 mm wide 'replacement' wall in the model is taken to be 0.22 W/m·K, otherwise it is taken to be 1.36 W/m·K.

Figure 4.8.2b shows the same 3D model in Figure 4.8.2a, but viewed from the opposite viewpoint, and shows the adiabatic boundaries (green) at the 'far' edges of the model at the location of the external wall[16] construction and the replacement 'adjacent' basement wall construction.

Figure 4.8.2c shows the 2D cross-section in the plane A–A' in Figure 4.8.2a of the 3D model of the junction and shows the various dimensions and boundary conditions that should be used in the 3D models. The height of the party wall (floor to ceiling) is h_{Pw} and the height of the basement wall (floor-to-ceiling) is h_{Bw}, which for both party wall and basement wall is fixed at 2.4 m. The characteristic dimension of the floor is b and is fixed at 8 m. The basement wall may be wider than the party wall above it, in which case the overlap of the basement wall with the heat-loss floor above is $W_o = (W_{Bw} - W_{Pw})$. The solid black line indicates where the floor slab and basement wall connect to the ground material and the dashed lines indicate the location of the external wall, which in the 2D cross-section of Figure 4.8.2b is 'replaced' by the vertical and horizontal adiabatic (green lines). These two adiabatic boundaries are applied to each of the adjacent floor edges of the model that are in contact with the ground (see Figure 4.8.2b).

The coloured lines in Figure 4.8.2c indicate the various boundary conditions with the detail given in the legend. The boundary condition in the legend marked '20 °C' and 'B_F' is that of the basement floor.

The first 3D model is of the complete junction and the surrounding ground as shown in Figure 4.8.2a. This 3D model gives the total heat flow through the model, Q_1^{3D}. Next, a second 3D model is created from the first in which the party wall and intermediate floor are removed and replaced by adiabatic boundaries (as shown in Figure 4.8.2d), where the vertical

---

[16] *Note:* the external wall and its foundation in the ground are not included in the model.

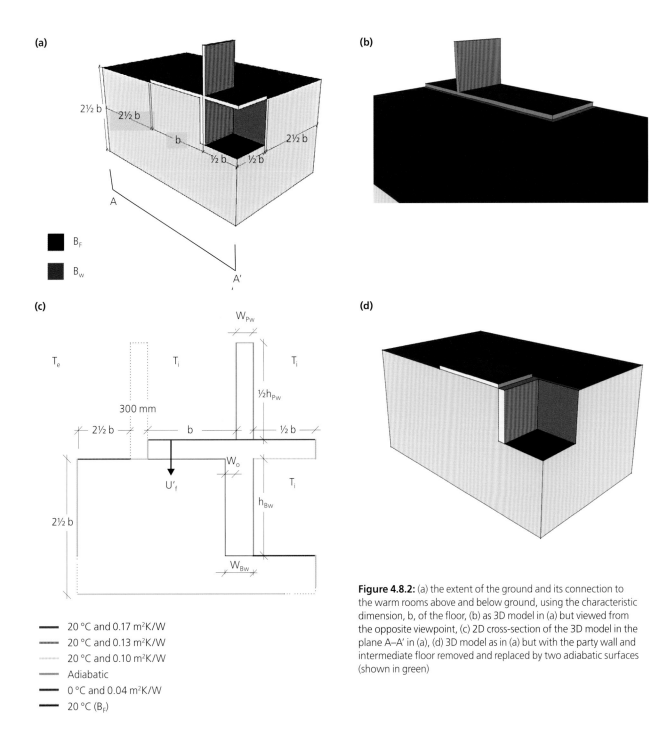

(a)

2½ b
2½ b
b
2½ b
½ b   ½ b

A

B_F

B_w

A'

(c)

W_{Pw}

T_e T_i T_i

½h_{Pw}

300 mm

2½ b   b   ½ b

W_o

U'_f T_i

h_{Bw}

2½ b

W_{Bw}

— 20 °C and 0.17 m²K/W
— 20 °C and 0.13 m²K/W
···· 20 °C and 0.10 m²K/W
— Adiabatic
— 0 °C and 0.04 m²K/W
— 20 °C (B_F)

(b)

(d)

Figure 4.8.2: (a) the extent of the ground and its connection to the warm rooms above and below ground, using the characteristic dimension, b, of the floor, (b) as 3D model in (a) but viewed from the opposite viewpoint, (c) 2D cross-section of the 3D model in the plane A–A' in (a), (d) 3D model as in (a) but with the party wall and intermediate floor removed and replaced by two adiabatic surfaces (shown in green)

adiabatic boundary is through the thickness of the heat-loss floor and the horizontal adiabatic boundary is through the full thickness of the basement wall. The total heat flow through the second 3D model is Q_2^{3D}. *Note:* the length of the junction, ℓ, in both 3D models is equal to ½b which is fixed at 4 m (ie b = 8 m).

The Ψ-value for the junction is calculated from:

$$\Psi = \frac{\left(Q_1^{3D} - Q_2^{3D} + U'_f \times \left(W_o \times \ell \times (T_i - T_e)\right)\right)}{\ell \times (T_i - T_e)}$$ 4.8.2 (1)

where U'_f is the U-value of the heat-loss floor determined from:

$$U'_f = \frac{Q_f}{(b - W_o) \times \ell \times (T_i - T_e)}$$ 4.8.2 (2)

where Q_f equals Q_2^{3D}, the heat flow into the heat-loss floor of the second 3D model.

If the thickness of the basement wall (W_{Bw}) is not greater than the thickness of the party wall (W_{Pw}) then equation 4.8.2 (1) is simplified to:

$$\Psi = \frac{\left(Q_1^{3D} - Q_2^{3D}\right)}{\ell \times (T_i - T_e)}$$ 4.8.2 (3)

See *Worked example 10* in Appendix B.

4.8.3 Ground floor (inverted)

This junction is similar to the party wall–ground floor (inverted) (see section 4.8.2), but without a party wall present, thus the approach used is essentially the same as that used for junction 4.8.2.

The various overall dimensions and boundary conditions are shown in Figure 4.8.3. Figure 4.8.3a shows the extent of the ground and its connection to the warm rooms above, using the characteristic dimension, b, of the floor.

The coloured lines in Figure 4.8.3c indicate the various boundary conditions with their detail given in the legend in the Figure. The boundary condition in the legend marked 20 °C and 'B$_F$' is that of the basement floor. The structure of the real basement floor is replaced in the model by a single layer 'floor' 150 mm thick and the real 'adjacent' basement wall
(ie at right angles to the basement wall of the junction) is replaced by a single layer 'wall' 300 mm wide. If the real basement floor is insulated then the thermal conductivity of the 150 mm thick 'replacement' floor in the model is taken to be 0.11 W/m·K, otherwise it is taken to be 1.88 W/m·K. If the

real 'adjacent' basement wall is insulated then the thermal conductivity of the 300 mm wide 'replacement' wall in the model is taken to be 0.22 W/m·K, otherwise it is taken to be 1.36 W/m·K. *Note:* the horizontal adiabatic is through the full thickness of the basement wall of the junction.

The first 3D model is of the complete junction and the surrounding ground as shown in Figure 4.8.3a. This 3D model gives the total heat flow through the model, Q_1^{3D}. Next, a second 3D model is created from the first in which the intermediate floor is removed and replaced by adiabatic boundaries as shown in Figure 4.8.3d, where the vertical adiabatic boundary is through the thickness of the heat-loss floor and the horizontal adiabatic is through the thickness of the basement wall. The total heat flow through the second 3D model is Q_2^{3D}.
Note: the length of the junction, ℓ, in both 3D models is equal to ½b which is fixed at 4 m.

The Ψ-value for the junction is calculated from:

$$\Psi = \frac{\left(Q_1^{3D} - Q_2^{3D}\right)}{\ell \times (T_i - T_e)} \qquad\qquad 4.8.3$$

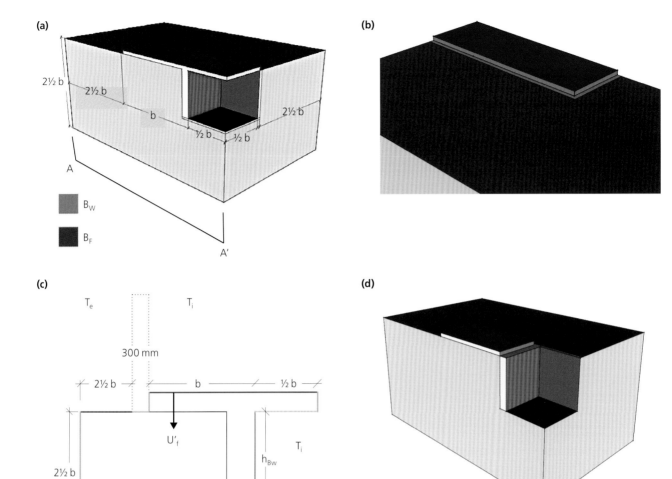

(a)

2½ b
2½ b
b
½ b   ½ b
2½ b

A

B_W

B_F

A'

(b)

(c)

T_e T_i

300 mm

2½ b   b   ½ b

U'_f

T_i

h_{Bw}

2½ b

W_{Bw}

—— 20 °C and 0.17 m²K/W
—— 20 °C and 0.13 m²K/W
········ 20 °C and 0.10 m²K/W
—— Adiabatic
—— 0 °C and 0.04 m²K/W
—— 20 °C and 0.17 m²K/W

(d)

Figure 4.8.3: (a) 3D model (model 1) of the junction showing the extent of the ground and its connection to the warm 'rooms', (b) as 3D model in (a) but viewed from the opposite viewpoint, (c) 2D cross-section of the 3D model in the plane A-A' in (a), (d) 3D model (model 2) as in (a) but with the Intermediate floor removed and replaced with two adiabatic surfaces (shown in green)

5 Point thermal bridges, χ-values

Low surface temperatures and increased heat flow also result from individual penetrations of the insulation layer by, for example, metal brackets that penetrate the structure to support a part of the building element such as a canopy or some other feature on the outside (or inside) of the building. Such point thermal bridges should be numerically modelled in 3D and a point thermal transmittance, χ, calculated (see 9.2 of BS EN ISO 10211:2007[6]).

The χ-value represents the extra heat flow through the 3D junction over and above that through any adjoining plane elements or linear thermal bridges. From the 3D numerical model that contains the point penetration, L^{3D}, is the thermal coupling coefficient between the internal and external environments and is calculated from:

$$L^{3D} = \frac{Q}{T_i - T_e} \quad (\text{W/K}) \tag{9}$$

where:

Q = total heat flow from the internal to the external environment (W)

T_i and T_e = temperatures of the internal and external environments (°C).

Hence the point thermal transmittance, χ, of the 3D junction is the residual heat flow from the internal to external environment after subtracting the one-dimensional heat flow through all flanking elements and the 2D heat flow through all 2D junctions, and is determined from:

$$\chi = L^{3D} - \sum (U \times A) - \sum \Psi \times \ell \quad (\text{W/K}) \tag{10}$$

where:

L^{3D} = thermal coupling coefficient
U = U-value (W/m²K) of the flanking element
A = area (m²) over which U applies
Ψ = linear thermal transmittance of the linear thermal bridge
ℓ = length (m) over which Ψ applies.

Where these point thermal bridges are repeating bridges in the plane flanking element, the additional heat loss should be incorporated into the calculation of the U-value of the plane building element that contains them, in which case a correction term, ΔU, is added to the calculation of the U-value, where:

$$\Delta U = \sum (n \times \chi) \quad (\text{W/m}^2\text{K}) \tag{11}$$

and where:

n = number of penetrations per m² in the façade of the building element that contains the point thermal bridges
χ = point thermal transmittance of the penetration (W/K).

In the case of single (non-repeating) penetrations, the resulting additional heat flow is added to the heat transmission factor, H, where:

$$H = \sum (A \times U) + \sum (\ell \times \Psi) + \sum \chi \quad (\text{W/K}) \tag{12}$$

6 Reporting of calculations

The two essential thermal performance values from the modelling of a junction are Ψ and f. The Ψ-value should be reported to three decimal places or three significant figures, whichever gives the greater precision, and the temperature factor, f, should be reported to two decimal places. In the case of disputed results, all modelling input and resulting output, as given in Appendix A, should be made available to an independent, third-party numerical modeller for adjudication.

7 References

1. HM Government. The Building Regulations 2010 (England). Approved Document L1A: Conservation of fuel and power in new dwellings, 2013 edn. Approved Document L1B: Conservation of fuel and power in existing dwellings, 2010 edn incorporating 2010, 2011 and 2013 amendments. Approved Document L2A: Conservation of fuel and power in new buildings other than dwellings, 2013 edn. Approved Document L2B: Conservation of fuel and power in existing buildings other than dwellings, 2010 edn incorporating 2010, 2011 and 2013 amendments. Available at: www.gov.uk.

2. Welsh Government. The Building Regulations 2010 (Wales). Approved Document L1A: Conservation of fuel and power in new dwellings, 2014 edn. Approved Document L1B: Conservation of fuel and power in existing dwellings, 2014 edn. Approved Document L2A: Conservation of fuel and power in new buildings other than dwellings, 2014 edn. Approved Document L2B: Conservation of fuel and power in existing buildings other than dwellings, 2014 edn. Available at: http://gov.wales.

3. Scottish Government. The Building (Scotland) Regulations 2004. Technical Handbooks: Domestic and Non-Domestic, 2015 edn. Section 6: Energy. Available at: www.gov.scot.

4. Northern Ireland Department of Finance and Personnel. The Building Regulations (Northern Ireland) 2012. Technical Booklet F1: Conservation of fuel and power in dwellings, 2012 edn incorporating February 2014 amendments. Technical Booklet F2: Conservation of fuel and power in buildings other than dwellings, 2012 edn incorporating February 2014 amendments. Available at: www.dfpni.gov.uk.

5. BSI. Building components and building elements. Thermal resistance and thermal transmittance. Calculation method. BS EN ISO 6946:2007. London, BSI, 2007.

6. BSI. Thermal bridges in building construction. Heat flows and surface temperatures. Detailed calculations. BS EN ISO 10211:2007. London, BSI, 2007.

7. Ward T I. Assessing the effect of thermal bridging at junctions and around openings. BRE IP 1/06. Bracknell, IHS BRE Press, 2006.

8. BSI. Thermal performance of buildings. Heat transfer via the ground. Calculation methods. BS EN ISO 13370: 2007. London, BSI, 2007.

9. BRE. U-value calculator. For information visit: http://projects.bre.co.uk/uvalues/.

10. Anderson B. Conventions for U-value calculations. BRE BR 443. Bracknell, IHS BRE Press, 2006.

11. BSI. Thermal performance of windows, doors and shutters. Calculation of thermal transmittance. BS EN ISO 10077. Part 1:2006 General. Part 2:2012 Numerical methods for frames. London, BSI.

12. BSI. Hygrothermal performance of building components and building elements. Internal surface temperature to avoid critical surface humidity and interstitial condensation. Calculation methods. BS EN ISO 13788:2012. London, BSI, 2012.

Appendices

Appendix A
Detailed input and output from a numerical model

This appendix gives the detailed input and output required from the model to facilitate any adjudication on the correct values of Ψ and f in a case of dispute. The input and output listed below is the minimum that should be kept for each model that provides values of Ψ and f. The input data file should also be kept, together with the name and version of the software that produced the modelling results.

Input data

1. Image(s) of the model showing the geometry

2. A list of the boundary conditions: temperature and surface conductance/resistance

3. A list of the materials and thermal conductivities

4. A list of air spaces, their dimensions and their thermal resistance (or equivalent thermal conductivity)

5. Either the total number of cells or the total number of nodes in the model

6. Image(s) giving an indication of the meshing used for the model, especially in the vicinity of the thermal bridge junction itself

7. The dimension of the flanking elements in each model, ie the distance from the junction to the adiabatic of the flanking element

Output data

1. Total heat flow into and out of the model

2. U-values of the flanking elements. If any of these are separate 1D calculations, then give the detail of the calculation. If instead it is determined from the heat flow or surface temperature at the adiabatic edge of the flanking element, that heat flow or temperature should be given

3. Image(s) showing internal surface temperatures

4. For 2D models, an image showing heat-flow lines through the model and the heat-flow value represented by each line

5. The detailed calculation of the Ψ-value, using the total heat flow through the model less the 1D heat flow through the flanking elements as determined from the product of area (or length) of the flanking elements. For each flanking element, give the U-value and area (or length) involved in the overall calculation of Ψ.

For adjudication, all minimum temperature factors should be available to three decimal places and all Ψ-values to three decimal places or three significant figures, whichever gives the greater precision.

Similarly, all U-values (with the exception of the U-value of a ground floor) that are used as input data for the model or used in the determination of Ψ, should be provided to three decimal places or three significant figures, whichever gives the greater precision, and the construction in the model should reflect this (modelling) U-value, U' (see section 3.1.3).

The U-value of a ground floor should be provided to four decimal places or four significant figures, whichever gives the greater precision.

All flanking dimensions should be given in metres (m) and should be provided to three decimal places.

Appendix B
Worked examples with calculated values of Ψ and f

This appendix contains 10 thermal bridge junctions where these are modelled and presented as worked examples showing the determination of their linear thermal transmittance, Ψ, and temperature factor, f. Each example covers a different junction type and these are listed in Table B.1. When determining Ψ and f from modelling each of these 10 junction types, the values obtained should agree with the Declared values listed within a tolerance of ± 0.01 W/m·K for Ψ and 0.01 for f.

In all of the examples modelled, the inside room temperature is taken to be 20 °C and the external temperature is taken to be 0 °C. For simplicity, these are the usual temperatures to use when modelling, but other temperatures can of course be used. For example, other internal and external temperatures can be helpful when focusing on the temperature profiles through the model under selected internal and external conditions.

Note: the Ψ-value and temperature factor, f, are not affected by a different choice for the internal and external temperatures; however, any temperatures used for intermediate spaces, such as the heat balance temperature T_U of an underfloor space in a suspended ground floor, would require to be re-calculated for the different set of internal and external temperatures selected for the particular model.

Table B.1: Worked examples

Worked example	Junction type	Description	SAP reference*
1	4.1.1	Eaves (pitched roof insulated at ceiling)	E10
2	4.2.3	Roof (sloping rafter to flat ceiling)	R6
3	4.3.1	Steel box-lintel (with perforated base-plate)	E1
4	4.4.3	Staggered party wall (horizontal)	E25
5	4.5.2	Balcony (penetrating wall insulation)	E23
6	4.6.2	Exposed floor (inverted)	E21
7.1	4.7.2	Suspended ground floor (with external wall) – concrete beams parallel to the junction but not included in the model	E5
7.2	4.7.2	Suspended ground floor (with external wall) – concrete beams perpendicular to the junction	E5
8	4.7.3	Basement wall/floor junction	E22
9	4.8.1	Party wall/ground-floor junction	P1
10	4.8.2	Party wall/ground floor (inverted) junction	P6

* This is the reference for the particular junction type as given in Table K1 of SAP 2012 (for more information, visit http://www.bre.co.uk/sap2012/page.jsp?id=2759).

Worked example 1

E10: Eaves (pitched roof insulated at ceiling)

The junction drawing shown in Figure B1 provides the information required to undertake the thermal assessment of the junction detail.

Step 1: Identify junction type
- Eaves (insulated at ceiling level)

Step 2: Refer to appropriate convention(s) within BR 497
- 4.1 Roof junctions
  - 4.1.1 Roof eaves (insulated at ceiling level)

Step 3: Define boundary conditions (see Table B1)
- As per section 2.5 *Surface heat transfer (surface resistances)*

Note: with the insulation of the roof at ceiling level (resulting in a ventilated cold roof space) the sloping roof construction and the eaves box are not included in the model.

Step 4: Apply appropriate formula

$$\Psi = \frac{Q^{2D} - U'_W \times \ell_W \times (T_i - T_e) - U'_C \times \ell_C \times (T_i - T_L)}{(T_i - T_e)}$$

where:

Q^{2D} = total heat flow through the 2D model (W/m)

U'_W = U-value of the wall (W/m²K)

U'_C = U-value of ceiling (W/m²K)

ℓ_W = length over which the wall U-value applies (m)

ℓ_C = length over which the ceiling U-value applies (m)

T_i = internal temperature (°C)

T_e = external temperature (°C)

T_L = loft-space temperature (°C)

Table B1: Boundary conditions in the model

Internal horizontal* heat flow	20.0 °C 0.13 m²K/W	
Internal upwards heat flow	20.0 °C 0.10 m²K/W	
External sheltered upwards heat flow (loft space/ ventilated areas)	0.0 °C 0.10 m²K/W	
External sheltered horizontal* heat flow	0.0 °C 0.13 m²K/W	
External exposed	0.0 °C 0.04 m²K/W	

* The values under 'horizontal' apply to heat flow directions ± 30° from the horizontal plane (eg if a roof slope is greater than 60°, the horizontal values should be used, otherwise the upwards (or downwards) values are used).

Step 5: Undertake calculations

$$\Psi = \frac{13.4813 - 0.268 \times 1.5 \times (20.0 - 0.0) - 0.143 \times 1.5 \times (20.0 - 0.0)}{(20.0 - 0.0)}$$

$$= 0.0576$$

Using equation (7):

$$f_{Rsi} = \frac{T_{si} - T_e}{T_i - T_e}$$

$$= \frac{17.67 - 0.0}{20.0 - 0.0} = 0.884$$

Step 6: Declare values

Ψ = 0.058 W/m·K
f_{Rsi} = 0.88

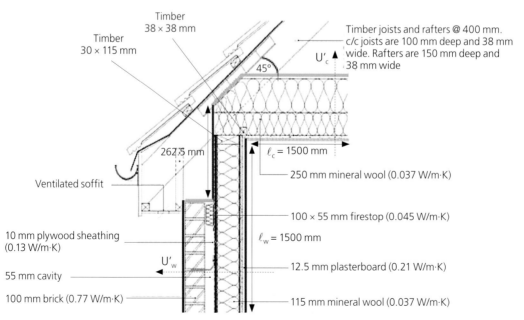

Timber 38 × 38 mm

Timber 30 × 115 mm

Timber joists and rafters @ 400 mm. c/c joists are 100 mm deep and 38 mm wide. Rafters are 150 mm deep and 38 mm wide

U'_C

45°

262.5 mm

ℓ_C = 1500 mm

Ventilated soffit

250 mm mineral wool (0.037 W/m·K)

100 × 55 mm firestop (0.045 W/m·K)

ℓ_W = 1500 mm

10 mm plywood sheathing (0.13 W/m·K)

U'_W

12.5 mm plasterboard (0.21 W/m·K)

55 mm cavity

100 mm brick (0.77 W/m·K)

115 mm mineral wool (0.037 W/m·K)

Figure B1: Worked example 1 – E10: Eaves (pitched roof insulated at ceiling)

Worked example 2

R6: Roof (sloping rafter to flat ceiling)

The junction drawing shown in Figure B2 provides the information required to undertake the thermal assessment of the junction detail.

Step 1: Identify junction type
- Roof to wall (flat ceiling: room-in-roof)

Step 2: Refer to appropriate convention(s) within BR 497
- 4.2 Room-in-roof junctions
  - 4.2.3 Sloping rafter to flat ceiling (insulated at ceiling)

Step 3: Define boundary conditions (see Table B2)
- As per section 2.5 *Surface heat transfer (surface resistances)*

Step 4: Apply appropriate formula

$$\Psi = \frac{Q^{2D} - U'_R \times \ell_R \times (T_i - T_e) - U'_C \times \ell_C \times (T_i - T_e)}{(T_i - T_e)}$$

where:

Q^{2D}	=	total heat flow through 2D model (W/m)
U'_R	=	U-value of wall (W/m²K)
U'_C	=	U-value of ceiling (W/m²K)
ℓ_R	=	length over which roof U-value applies (m)
ℓ_C	=	length over which ceiling U-value applies (m)
T_i	=	internal temperature (°C)
T_e	=	external temperature (°C)

Table B2: Boundary conditions in the model

Internal upwards heat flow*	20.0 °C	
	0.10 m² K/W	
External sheltered upwards heat flow (loft space/ ventilated areas)*	0.0 °C	
	0.10 m² K/W	

* The values under 'horizontal' apply to heat flow directions ± 30° from the horizontal plane (eg if a roof slope is greater than 60°, the horizontal values should be used, otherwise the upwards (or downwards) values are used).

Step 5: Undertake calculation

$$\Psi = \frac{7.0894 - 0.103 \times 1.5 \times (20.0 - 0.0) - 0.108 \times 1.5 \times (20.0 - 0.0)}{(20.0 - 0.0)}$$

$$= 0.0380$$

Using equation (7):

$$f_{Rsi} = \frac{19.45 - 0.0}{20.0 - 0.0} = 0.973$$

Step 6: Declare values

Ψ = 0.038 W/m·K
f_{Rsi} = 0.97

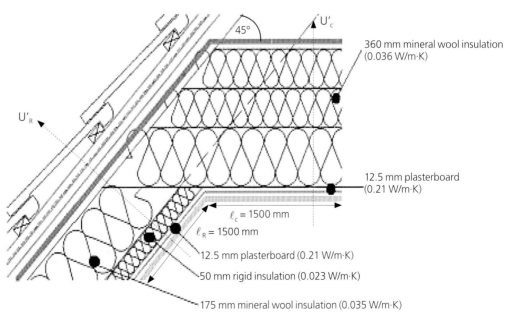

45°

U'_C

360 mm mineral wool insulation (0.036 W/m·K)

U'_R

12.5 mm plasterboard (0.21 W/m·K)

ℓ_C = 1500 mm

ℓ_R = 1500 mm

12.5 mm plasterboard (0.21 W/m·K)

50 mm rigid insulation (0.023 W/m·K)

175 mm mineral wool insulation (0.035 W/m·K)

Figure B2: Worked example 2 – R6: Roof (sloping rafter to flat ceiling)

Worked example 3

E1: Steel lintel with perforated steel base-plate

The junction drawing shown in Figure B3a provides the information required to undertake the thermal assessment of the junction detail.

Step 1: Identify junction type
- Lintel, with steel perforated base-plate

Step 2: Refer to appropriate convention(s) within BR 497
- 4.3 (Junctions around openings)
  - 4.3.1 Lintel

Step 3: Define boundary conditions (see Table B3a)
- As per section 2.5 *Surface heat transfer (surface resistances)*

Step 4: Review cross-section of steel box lintel (Figure B3.2; *note:* drawing is not to scale) and calculate thermal conductivity of lintel air spaces
For the air spaces adjacent to and within the box lintel (marked 1, 2 and 3 in Figure B3b), Table B3b gives the maximum dimensions of the actual air spaces, where D is in, and B parallel to, the principal heat flow direction, and where d and b are the dimensions of the transformed rectangular air spaces in accordance with section 2.4.3. The thermal resistance of each transformed air space is then calculated using the equations in Annex B of BS EN ISO 6946[5] for divided air spaces.

Step 5: Calculate equivalent thermal conductivity of the perforated base-plate from a separate model (see section 2.3.1)
Figure B3c shows the plan view of the perforated base of the lintel and the dimensions of the repeat perforation. The slotted

Table B3a: Boundary conditions in the model

Internal horizontal heat flow	20.0 °C 0.13 m²K/W	
Internal upwards heat flow	20.0 °C 0.10 m²K/W	
Adiabatic	0 m²K/W	
External exposed	0.0 °C 0.04 m²K/W	

holes are staggered perpendicular to the heat flow direction, with a slot size of 60 mm × 10 mm with semi-circular ends that have a radius of 5 mm. The slots are filled with plaster with a thermal conductivity of 0.57 W/m·K. Refer to section 2.3.1 *Perforated metal plates* for appropriate boundary conditions. *Note:* the thermal conductivity of the steel of the base-plate is 50 W/m·K.

From the numerical modelling of the perforated base-plate as described above, the equivalent thermal conductivity of the base-plate should be 6.9 ± 0.10 W/m·K.

Step 6: Apply equivalent thermal conductivity of the lintel air spaces and of the perforated base-plate to the numerical model of the lintel junction detail

Step 7: Apply appropriate formula

$$\Psi = \frac{Q^{2D} - U'_w \times \ell_w \times (T_i - T_e)}{(T_i - T_e)}$$

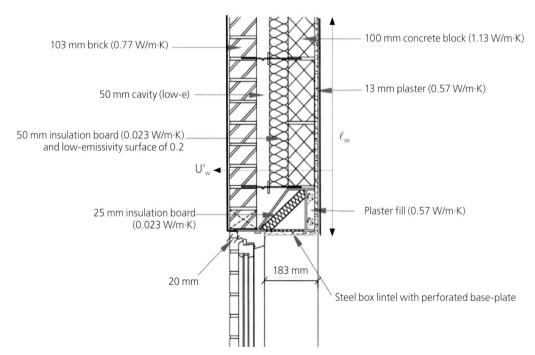

103 mm brick (0.77 W/m·K)

50 mm cavity (low-e)

50 mm insulation board (0.023 W/m·K) and low-emissivity surface of 0.2

U'_w

25 mm insulation board (0.023 W/m·K)

20 mm

183 mm

100 mm concrete block (1.13 W/m·K)

13 mm plaster (0.57 W/m·K)

ℓ_w

Plaster fill (0.57 W/m·K)

Steel box lintel with perforated base-plate

Figure B3a: Worked example 3 – Vertical cross-section through the lintel junction

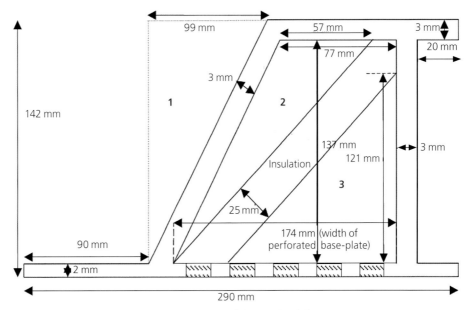

Figure B3b: Worked example 3 – Dimensions for the vertical cross-section of the steel lintel showing air spaces

Table B3b: Calculation of equivalent thermal conductivity of lintel air spaces

Air space	D (mm)	B (mm)	Area (mm²)	d (mm)	b (mm)	Resistance (m²K/W)	Conductivity (W/m·K)
1	99	140	6930	70	99	0.218	0.322
2	154	137	3905	66	59	0.231	0.286
3	137	121	8289	97	86	0.231	0.420

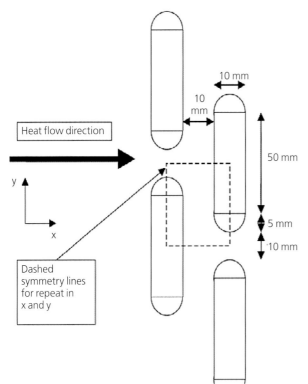

where:

Q^{2D} = total heat flow through 2D model (W/m)

U'_w = U-value of wall (W/m²K)

ℓ_w = length over which wall U-value applies (m)

T_i = internal temperature (°C)

T_e = external temperature (°C)

Step 8: Undertake calculation

$$\Psi = \frac{18.2149 - 0.330 \times 1.5 \times (20.0 - 0.0)}{(20.0 - 0.0)}$$

$$= 0.416$$

- Using equation (7):

$$f_{Rsi} = \frac{15.08 - 0.0}{20.0 - 0.0} = 0.754$$

Step 9: Declare values

Ψ = 0.42 W/m·K

f_{Rsi} = 0.75

Figure B3c: Worked example 3 – Slot size and base-plate perforation repeat

Worked example 4

E25: Staggered party wall between dwellings

The junction drawing shown in Figure B4 provides the information required to undertake the thermal assessment of the junction detail.

Step 1: Identify junction type
- Staggered party wall between dwellings

Step 2: Refer to appropriate convention(s) within BR 497
- 4.4 Corner junctions
  - 4.4.3 Staggered party wall (horizontal)

Step 3: Define boundary conditions (see Table B4)
- As per section 2.5 *Surface heat transfer (surface resistances)*

Step 4: Apply appropriate formula

$$\Psi = \frac{Q^{2D} - U'_{W_1} \times \ell_{W_1} \times (T_i - T_e) - U'_{W_2} \times \ell_{W_2} \times (T_i - T_e)}{(T_i - T_e)}$$

where:

Q^{2D} = total heat flow through 2D model (W/m)

U'_{W_1} = U-value of wall 1 (W/m²K)

ℓ_{W_1} = length over which wall 1 U-value applies (m)

U'_{W_2} = U-value of wall 2 (W/m²K)

ℓ_{W_2} = length over which wall 2 U-value applies (m)

T_i = internal temperature (°C)

T_e = external temperature (°C)

Table B4: Boundary conditions in the model

Internal horizontal heat flow	20.0 °C 0.13 m²K/W	
External boundary	0.0 °C 0.04 m²K/W	

Step 5: Undertake calculation

$$\Psi = \frac{11.3916 - 0.156 \times 1.5 \times (20.0 - 0.0) - 0.156 \times 1.524 \times (20.0_i - 0.0)}{(20.0 - 0.0)}$$

$$= 0.094$$

Using equation (7):

$$f_{Rsi} = \frac{19.15 - 0.0}{20.0 - 0.0} = 0.958$$

Step 6: Declare values

Ψ = 0.094 W/m·K
f_{Rsi} = 0.96

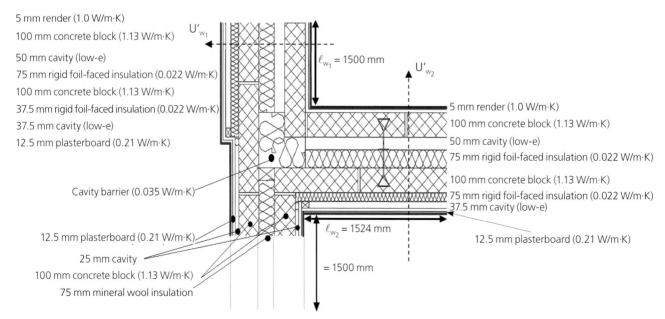

5 mm render (1.0 W/m·K)
100 mm concrete block (1.13 W/m·K)
50 mm cavity (low-e)
75 mm rigid foil-faced insulation (0.022 W/m·K)
100 mm concrete block (1.13 W/m·K)
37.5 mm rigid foil-faced insulation (0.022 W/m·K)
37.5 mm cavity (low-e)
12.5 mm plasterboard (0.21 W/m·K)

Cavity barrier (0.035 W/m·K)

12.5 mm plasterboard (0.21 W/m·K)
25 mm cavity
100 mm concrete block (1.13 W/m·K)
75 mm mineral wool insulation

U'_{W_1}
ℓ_{W_1} = 1500 mm
U'_{W_2}

5 mm render (1.0 W/m·K)
100 mm concrete block (1.13 W/m·K)
50 mm cavity (low-e)
75 mm rigid foil-faced insulation (0.022 W/m·K)
100 mm concrete block (1.13 W/m·K)
75 mm rigid foil-faced insulation (0.022 W/m·K)
37.5 mm cavity (low-e)

ℓ_{W_2} = 1524 mm
= 1500 mm

12.5 mm plasterboard (0.21 W/m·K)

Figure B4: Worked example 4 – E25 Staggered party wall (horizontal section)

Worked example 5

E23: Balcony between dwellings, balcony support penetrates wall insulation

The junction drawing shown in Figure B5 provides the information required to undertake the thermal assessment of the junction detail.

Step 1: Identify junction type
- Balcony

Step 2: Refer to appropriate convention(s) within BR 497
- 4.5 Intermediate-floor or party-wall junctions (with external wall)
  - 4.5.2 Balcony

Step 3: Define boundary conditions (see Table B5)
- As per section 2.5 *Surface heat transfer (surface resistances)*

Step 4: Apply appropriate formula

$$\Psi = \frac{Q^{2D} - U'_W \times \ell_W \times (T_i - T_e)}{(T_i - T_e)}$$

where:

Q^{2D} = total heat flow through 2D model (W/m)

U'_W = U-value of wall (W/m²K)

ℓ_W = length over which wall U-value applies (m)
where $\ell_W = a + c^*$ with half of Ψ allocated to each dwelling

---

* Where the balcony is within a single dwelling, $\ell_W = a + b + c$ and the full Ψ-value is allocated to the dwelling.

Table B5: Boundary conditions in the model

Internal horizontal heat flow	20.0 °C 0.13 m²K/W	
Internal upwards heat flow	20.0 °C 0.10 m²K/W	
Internal downwards heat flow	20.0 °C 0.10 m²K/W	
External boundary	0.0 °C 0.04 m²K/W	

T_i = internal temperature (°C)

T_e = external temperature (°C)

Step 5: Undertake calculation

$$\Psi = \frac{43.7564 - 0.361 \times 3.0 \times (20.0 - 0.0)}{(20.0 - 0.0)}$$

$$= 1.105$$

Using equation (7):

$$f_{Rsi} = \frac{19.15 - 0.0}{20.0 - 0.0} = 0.958$$

Step 6: Declare values

Ψ = 1.105 W/m·K (with half of this value allocated to each dwelling)

f_{Rsi} = 0.82

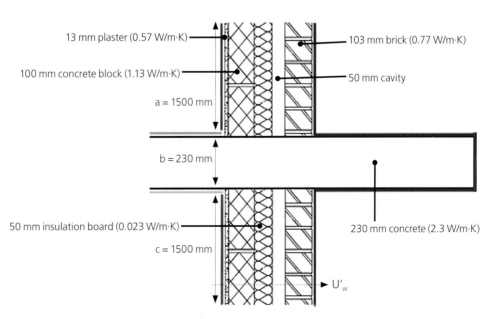

13 mm plaster (0.57 W/m·K)

100 mm concrete block (1.13 W/m·K)

a = 1500 mm

b = 230 mm

50 mm insulation board (0.023 W/m·K)

c = 1500 mm

103 mm brick (0.77 W/m·K)

50 mm cavity

230 mm concrete (2.3 W/m·K)

U'_W

Figure B5: Worked example 5 – E23 Balcony (penetrating wall insulation)

Worked example 6

E21: Exposed floor (inverted)

The junction drawing shown in Figure B6 provides the information required to undertake the thermal assessment of the junction detail.

Step 1: Identify junction type
- Exposed floor (inverted)

Step 2: Refer to appropriate convention(s) within BR 497
- 4.6 Exposed floors
  - 4.6.2 Exposed floor (inverted)

Step 3: Define boundary conditions (see Table B6)
As per section 2.5 *Surface heat transfer (surface resistances)*

Step 4: Apply appropriate formula

$$\Psi = \frac{Q^{2D} - U'_W \times \ell_W \times (T_i - T_e) - U'_F \times \ell_F \times (T_i - T_e)}{(T_i - T_e)}$$

where:

Q^{2D} = total heat flow through 2D model (W/m)

U'_W = U-value of wall (W/m²K)

ℓ_W = length over which U-value applies for wall (m)

U'_f = U-value of exposed floor (W/m²K)

ℓ_f = length over which U-value applies for exposed floor (m)

T_i = internal temperature (°C)

T_e = external temperature (°C)

Table B6: Boundary conditions in the model

Internal horizontal heat flow	20.0 °C 0.13 m²K/W	
Internal upwards heat flow	20.0 °C 0.10 m²K/W	
Internal downwards heat flow	20.0 °C 0.10 m²K/W	
External boundary	0.0 °C 0.04 m²K/W	

Step 5: Undertake calculation

$$\Psi = \frac{12.336 - 0.156 \times 1.7305 \times (20.0 - 0.0) - 0.159 \times 1.5 \times (20.0 - 0.0)}{(20.0 - 0.0)}$$

$$= 0.108$$

Using equation (7):

$$f_{Rsi} = \frac{18.71 - 0.0}{20.0 - 0.0} = 0.936$$

Step 6: Declare values

Ψ = 0.108 W/m·K
f_{Rsi} = 0.94

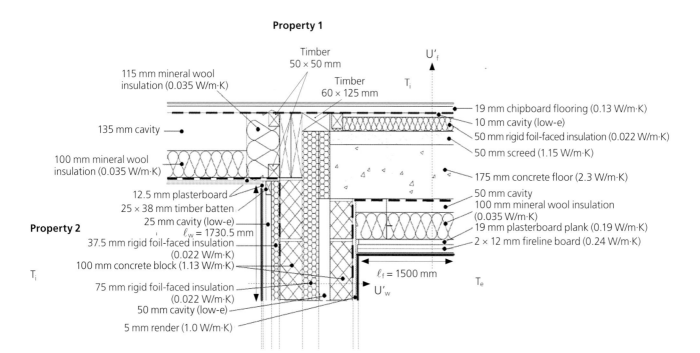

Property 1

- 115 mm mineral wool insulation (0.035 W/m·K)
- Timber 50 × 50 mm
- Timber 60 × 125 mm
- U'_f
- T_i
- 135 mm cavity
- 100 mm mineral wool insulation (0.035 W/m·K)
- 12.5 mm plasterboard
- 25 × 38 mm timber batten
- 25 mm cavity (low-e)
- ℓ_W = 1730.5 mm
- 37.5 mm rigid foil-faced insulation (0.022 W/m·K)
- 100 mm concrete block (1.13 W/m·K)
- 75 mm rigid foil-faced insulation (0.022 W/m·K)
- 50 mm cavity (low-e)
- 5 mm render (1.0 W/m·K)
- **Property 2**
- T_i
- 19 mm chipboard flooring (0.13 W/m·K)
- 10 mm cavity (low-e)
- 50 mm rigid foil-faced insulation (0.022 W/m·K)
- 50 mm screed (1.15 W/m·K)
- 175 mm concrete floor (2.3 W/m·K)
- 50 mm cavity
- 100 mm mineral wool insulation (0.035 W/m·K)
- 19 mm plasterboard plank (0.19 W/m·K)
- 2 × 12 mm fireline board (0.24 W/m·K)
- ℓ_f = 1500 mm
- T_e
- U'_w

Figure B6: Worked example 6 – E21 Exposed floor (inverted)

Worked example 7

E5: Suspended ground floor

The junction drawings shown in Figures B7a and B7b provide the information required to undertake the thermal assessment of two junction details:

- where the concrete beams are parallel to the junction
- where the concrete beams are perpendicular to the junction.

For each modelled junction:

Step 1: Identify junction type
- Concrete beam and block ground floor

Step 2: Refer to appropriate convention(s) within BR 497
- 4.7 Ground-floor junctions
    - 4.7.2 Suspended ground floor

Step 3: Define boundary conditions (see Table B7)
As per section 2.5 *Surface heat transfer (surface resistances)*

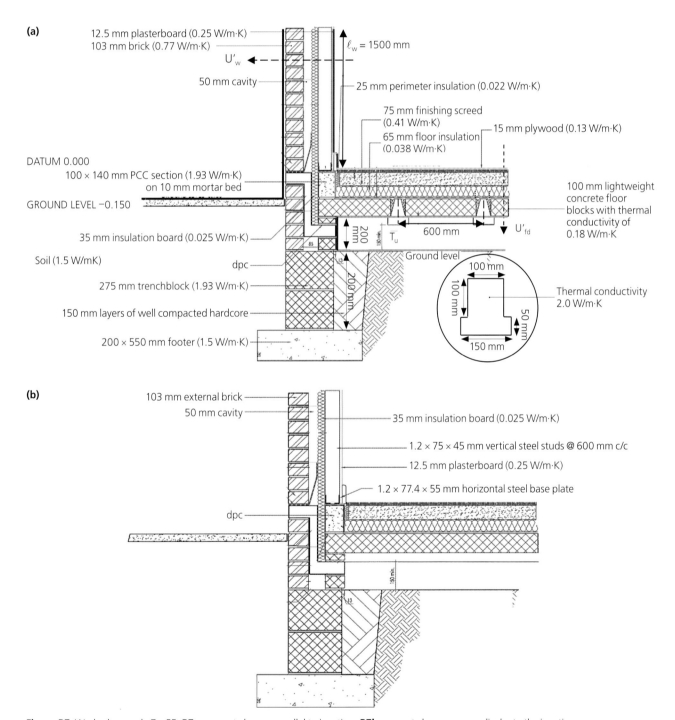

Figure B7: Worked example 7 – E5. **B7a:** concrete beams parallel to junction, **B7b:** concrete beams perpendicular to the junction (dimensions, material properties and boundary conditions are as in **a**)

Table B7: Boundary conditions in the model

Internal horizontal heat flow	20.0 °C 0.13 m²K/W
Internal downwards heat flow	20.0 °C 0.10 m²K/W
Underfloor space downwards heat flow	T_u °C* 0.10 m²K/W
Underfloor space horizontal heat flow	T_u °C* 0.13 m²K/W
External exposed	0.0 °C 0.04 m²K/W

* The underfloor space is at an intermediate temperature, T_u, between T_i and T_e. T_u is calculated from the heat balance in accordance with Annex E of BS EN ISO 13370[7].

For T_i = 20.0 °C and T_e = 0.0 °C, T_u for the suspended floor is calculated to be 6.12 °C.

Step 4: Apply appropriate formula

$$\Psi = \frac{Q^{2D} - U'_w \times \ell_w \times (T_i - T_e) - U'_f \times \tfrac{1}{2}b \times (T_i - T_e)}{(T_i - T_e)}$$

where:

Q^{2D} = total heat flow through 2D model (W/m)

U'_w = U-value of wall (W/m²k)

ℓ_w = length over which U-value applies (m)

U'_f = U-value of ground floor (W/m²k)

$\tfrac{1}{2}b$ = length over which the floor U-value applies, where b (the characteristic dimension of the floor from BS EN ISO 13370[7]) is fixed at 8 m; hence $\tfrac{1}{2}b$ in the model is set at 4 m

T_i = internal temperature (°C)

T_e = external temperature (°C)

Step 5: Undertake calculations

- *Concrete beam parallel to junction*

$$\Psi = \frac{35.2485 - 0.472 \times 1.5 \times (20.0 - 0.0) - 0.3443 \times 4.0 \times (20.0 - 0.0)}{(20.0 - 0.0)}$$

$$= 0.099$$

Using equation (7):

$$f_{Rsi} = \frac{16.79 - 0.0}{20.0 - 0.0} = 0.840$$

- *Concrete beam perpendicular to junction*

$$\Psi = \frac{36.9053 - 0.472 \times 1.5 \times (20.0 - 0.0) - 0.3674 \times 4.0 \times (20.0 - 0.0)}{(20.0 - 0.0)}$$

$$= 0.118$$

Using equation (7):

$$f_{Rsi} = \frac{16.35 - 0.0}{20.0 - 0.0} = 0.818$$

- *Ground-floor corner*

Using equation (7):

$$f_{Rsi} = \frac{15.3 - 0.0}{20.0 - 0.0} = 0.765$$

Step 6: Declare values

- *Concrete beam parallel to junction*
Ψ = 0.099 W/m·K
f_{Rsi} = 0.84

- *Concrete beam perpendicular to junction*
Ψ = 0.118 W/m·K
f_{Rsi} = 0.82

- *Ground-floor corner* (see section 3.2.1)
$f_{Rsi\,(3D\,corner)}$ = 0.77

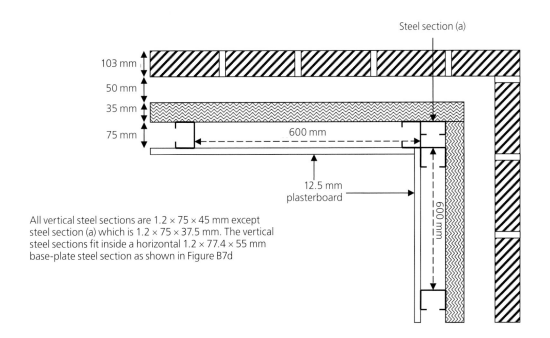

Figure B7c: Horizontal cross-section of wall corner showing steels and their dimensions and spacing

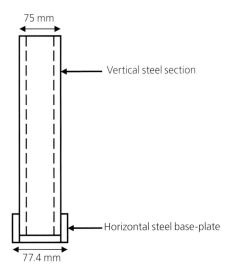

Figure B7d: Vertical cross-section showing dimensions and spacing of the vertical steel connecting to the horizontal steel base-plate

Worked example 8

E22: Basement wall/floor junction

The junction drawing shown in Figure B8 provides the information required to undertake the thermal assessment of the junction detail.

Step 1: Identify junction type
- Basement wall/floor junction

Step 2: Refer to appropriate convention(s) within BR 497
- 4.7 Ground-floor junctions
  - 4.7.3 Basement wall/floor junction

Step 3: Define boundary conditions (see Table B8)
- As per section 2.5 *Surface heat transfer (surface resistances)*

Step 4: Apply appropriate formula

$$\Psi = \frac{Q^{2D} - U'_{Bw} \times h_{Bw} \times (T_i - T_e) - U'_{Bf} \times \ell_{Bf} \times (T_i - T_e)}{(T_i - T_e)}$$

where:

Q^{2D} = total heat flow through 2D model (W/m)

U'_{Bw} = U-value of basement wall (W/m²K) (from calculation)

U'_{Bf} = U-value of basement floor (W/m²K) (from calculation)

h_{Bw} = height of basement wall over which U-value applies (2.4 m)

ℓ_{Bf} = length of basement floor over which U-value applies (4 m)*

b = 8 m

T_i = internal temperature (°C)

T_e = external temperature (°C)

Table B8: Boundary conditions in the model

Internal horizontal heat flow	20.0 °C 0.13 m²K/W	
Internal downwards heat flow	20.0 °C 0.17 m²K/W	
Adiabatic	0 m²K/W	
External exposed	0.0 °C 0.04 m²K/W	

Step 5: Undertake calculation

$$\Psi = \frac{16.4656 - 0.129 \times 2.4 \times (20.0 - 0.0) - 0.1161 \times 4.0 \times (20.0 - 0.0)}{(20.0 - 0.0)}$$

$$= 0.0486$$

Using equation (7):

$$f_{Rsi} = \frac{18.84 - 0.0}{20.0 - 0.0} = 0.942$$

Step 6: Declare values

Ψ = 0.048 W/m·K
f_{Rsi} = 0.94

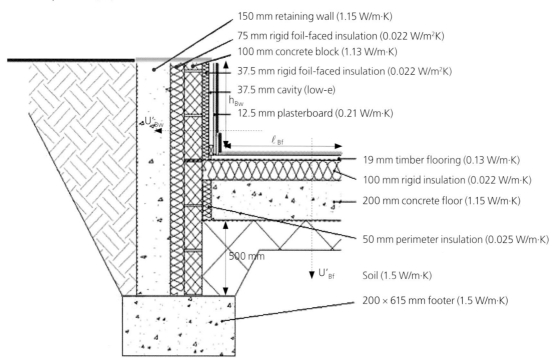

150 mm retaining wall (1.15 W/m·K)
75 mm rigid foil-faced insulation (0.022 W/m²K)
100 mm concrete block (1.13 W/m·K)
37.5 mm rigid foil-faced insulation (0.022 W/m²K)
37.5 mm cavity (low-e)
12.5 mm plasterboard (0.21 W/m·K)
19 mm timber flooring (0.13 W/m·K)
100 mm rigid insulation (0.022 W/m·K)
200 mm concrete floor (1.15 W/m·K)
50 mm perimeter insulation (0.025 W/m·K)
Soil (1.5 W/m·K)
200 × 615 mm footer (1.5 W/m·K)

Figure B8: Worked example 8 – E22 Basement wall/floor
* *Note:* ℓ_{Bf} is measured to the finished wall surface including the thickness of any skirting.

Worked example 9

P1: Party wall/ground-floor junction

The junction drawing shown in Figure B9 provides the information required to undertake the thermal assessment of the junction detail.

Step 1: Identify junction type
• Party wall/ground floor – between dwellings

Step 2: Refer to appropriate convention(s) within BR 497
• 4.8 Special-case junctions that connect to the ground
  – 4.8.1 Party wall/ground-floor junction

Step 3: Define boundary conditions (see Table B9)
• As per section 2.5 *Surface heat transfer (surface resistances)*

Step 4: Apply appropriate formula

$$\Psi_c = \frac{\left(Q_1^{3D} - Q_2^{3D}\right)}{(T_i - T_e) \times \ell}$$

Note: U'_f = U-value of the heat loss from the floor determined by:

$$U'_f = \frac{Q_f}{\left(b + \frac{1}{2}b + t_{wall}\right) \times \ell \times (T_i - T_e)}$$

where:

Q_1^{3D} = Total heat flow through 3D model 1 (W)

Q_2^{3D} = Total heat flow through 3D model 2 with party wall removed (W)

ℓ = length of party wall (4 m)

b = 8 m

t_{wall} = thickness of party wall

Table B9: Boundary conditions in the model

Internal horizontal heat flow	20.0 °C 0.13 m²K/W	
Internal downwards heat flow	20.0 °C 0.17 m²K/W	
External exposed	0.0 °C 0.04 m²K/W	

T_i = Internal temperature (°C)

T_e = External temperature (°C)

Step 5: Undertake calculation

$$U'_f = \frac{139.892}{(8.0 + 4.0 + 0.364) \times 4.0 \times (20.0 - 0.0)}$$

$$= 0.1414$$

$$\Psi_c = \frac{(142.0512 - 139.892)}{(20.0 - 0.0) \times 4.0} = 0.0270$$

$$\Psi = \Psi_c + (t_{wall} \times U'_f)$$

$$\Psi = 0.0270 + (0.364 \times 0.1414)$$

$$\Psi = 0.0785$$

Using equation (7):

$$f_{Rsi} = \frac{19.29 - 0.0}{20.0 - 0.0} = 0.964$$

Step 6: Declare values

Ψ = 0.079 W/m·K

f_{Rsi} = 0.96

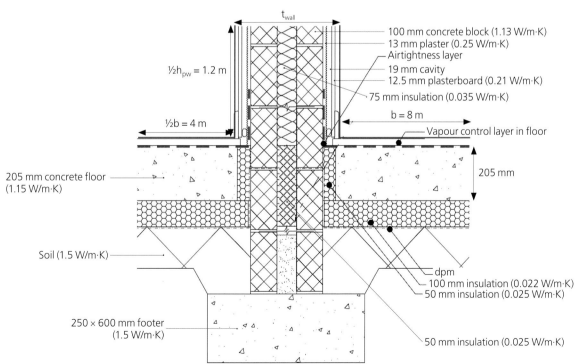

100 mm concrete block (1.13 W/m·K)
13 mm plaster (0.25 W/m·K)
Airtightness layer
19 mm cavity
12.5 mm plasterboard (0.21 W/m·K)
75 mm insulation (0.035 W/m·K)

½h_pw = 1.2 m

t_wall

½b = 4 m

b = 8 m

Vapour control layer in floor

205 mm concrete floor (1.15 W/m·K)

205 mm

Soil (1.5 W/m·K)

dpm
100 mm insulation (0.022 W/m·K)
50 mm insulation (0.025 W/m·K)

250 × 600 mm footer (1.5 W/m·K)

50 mm insulation (0.025 W/m·K)

Figure B9: Worked example 9 – P1 Party wall/ground floor

Worked example 10

P6: Party wall/ground-floor (inverted) junction

The junction drawing shown in Figure B10a provides the information required to undertake the thermal assessment of the junction detail (model 1). Figure B10b shows the location (solid green lines) of the horizontal and vertical adiabatic boundaries for model 2 that facilitates the removal from model 1 of the party wall and intermediate floor constructions (indicated by the area of green shading).

Step 1: Identify junction type
- Party wall/ground floor (inverted) – between dwellings

Step 2: Refer to appropriate convention(s) within BR 497
- 4.8 Special-case junctions that connect to the ground
  - 4.8.2 Party wall/ground-floor junction (inverted)

Step 3: Define boundary conditions (see Table B10)
- As per section 2.5 *Surface heat transfer (surface resistances)*

Step 4: Apply appropriate formula

$$\Psi = \frac{\left(Q_1^{3D} - Q_2^{3D} - U_f \times \left(W_o \times \ell \times (T_i - T_e)\right)\right)}{(T_i - T_e) \times \ell}$$

Note: U_f = U-value of the heat loss from the floor determined by:

$$U_f = \frac{Q_f}{\ell \times (b - W_o) \times (T_i - T_e)}$$

where:

Q_1^{3D} = total heat flow through 3D model 1 (W)

Q_2^{3D} = total heat flow through 3D model 2 with adiabatic replacing party wall and intermediate floor (W)

Q_f = heat loss through the ground floor of model 2

ℓ = length of party wall (4 m)

Table B10: Boundary conditions in the model

Internal horizontal heat flow	20.0 °C 0.13 m²K/W	
Internal upwards heat flow	20.0 °C 0.10 m²K/W	
Internal downwards heat flow	20.0 °C 0.17 m²K/W	
External exposed	0.0 °C 0.04 m²K/W	

b = 8 m

W_o = overlap of the basement wall with the ground floor (0.0365 m)

T_i = internal temperature (°C)

T_e = external temperature (°C)

Step 5: Undertake calculation

$$U_f = \frac{81.8349}{4.0 \times (8.0 - 0.0365) \times (20.0 - 0.0)} = 0.1285$$

$$\Psi = \frac{(419.5903 - 394.8985 - 0.1285 \times (0.0365 \times 4.0 \times (20.0 - 0.0)))}{(20.0 - 0.0) \times 4.0}$$

$$= 0.3040$$

Using equation (7):

$$f_{Rsi} = \frac{18.61 - 0.0}{20.0 - 0.0} = 0.931$$

Step 6: Declare values

Ψ = 0.304 W/m·K
f_{Rsi} = 0.93

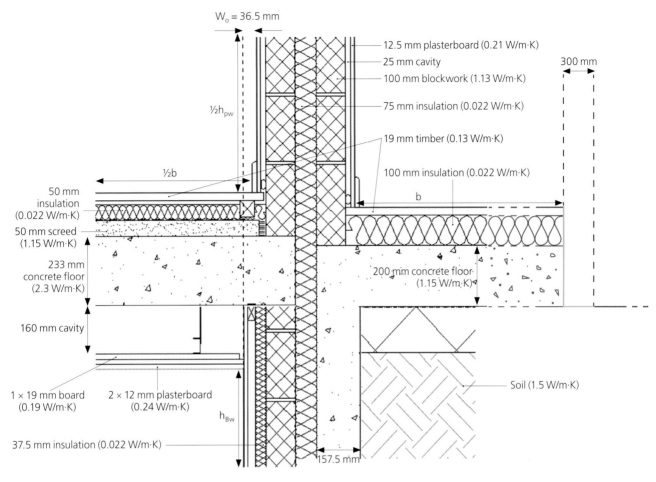

W$_o$ = 36.5 mm

12.5 mm plasterboard (0.21 W/m·K)

25 mm cavity

100 mm blockwork (1.13 W/m·K)

75 mm insulation (0.022 W/m·K)

19 mm timber (0.13 W/m·K)

100 mm insulation (0.022 W/m·K)

300 mm

½h$_{pw}$

½b

b

50 mm insulation (0.022 W/m·K)

50 mm screed (1.15 W/m·K)

233 mm concrete floor (2.3 W/m·K)

200 mm concrete floor (1.15 W/m·K)

160 mm cavity

1 × 19 mm board (0.19 W/m·K)

2 × 12 mm plasterboard (0.24 W/m·K)

Soil (1.5 W/m·K)

h$_{Bw}$

37.5 mm insulation (0.022 W/m·K)

157.5 mm

Figure B10a: Worked example 10 – P6 Party wall/ground floor (inverted): Model 1

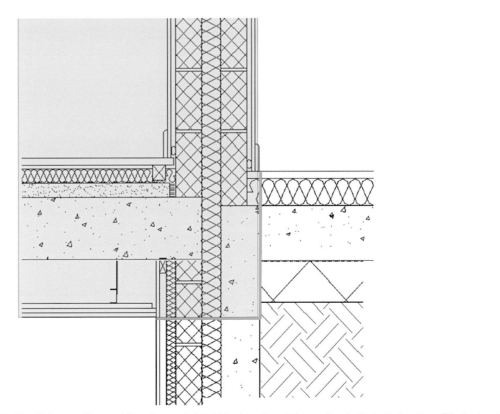

Figure B10b: Worked example 10 – P6 Party wall/ground floor (inverted): Model 2 where the solid green lines indicate the location of the horizontal and vertical adiabatic boundaries

Publications from IHS BRE Press

Fire performance of external thermal insulation for walls of multistorey buildings. 3rd edn. **BR 135**

External fire spread. 2nd edn. **BR 187**

Site layout planning for daylight and sunlight. 2nd edn. **BR 209**

Radon: guidance on protective measures for new buildings. 2015 edn. **BR 211**

Cracking in buildings. 2nd edn. **BR 292**

Fire safety and security in retail premises. **BR 508**

Automatic fire detection and alarm systems. **BR 510**

Handbook for the structural assessment of large panel system (LPS) dwelling blocks for accidental loading. **BR 511**

Ventilation for healthy buildings: reducing the impact of urban pollution. **FB 30**

Financing UK carbon reduction projects. **FB 31**

The cost of poor housing in Wales. **FB 32**

Dynamic comfort criteria for structures: a review of UK standards, codes and advisory documents. **FB 33**

Water mist fire protection in offices: experimental testing and development of a test protocol. **FB 34**

Airtightness in commercial and public buildings. 3rd edn. FB 35

Biomass energy. **FB 36**

Environmental impact of insulation. **FB 37**

Environmental impact of vertical cladding. **FB 38**

Environmental impact of floor finishes: incorporating The Green Guide ratings for floor finishes. **FB 39**

LED lighting. **FB 40**

Radon in the workplace. 2nd edn. **FB 41**

U-value conventions in practice. **FB 42**

Lessons learned from community-based microgeneration projects: the impact of renewable energy capital grant schemes. **FB 43**

Energy management in the built environment: a review of best practice. **FB 44**

The cost of poor housing in Northern Ireland. **FB 45**

Ninety years of housing, 1921–2011. **FB 46**

BREEAM and the Code for Sustainable Homes on the London 2012 Olympic Park. **FB 47**

Saving money, resources and carbon through SMARTWaste. **FB 48**

Concrete usage in the London 2012 Olympic Park and the Olympic and Paralympic Village and its embodied carbon content. **FB 49**

A guide to the use of urban timber. **FB 50**

Low flow water fittings: will people accept them? **FB 51**

Evacuating vulnerable and dependent people from buildings in an emergency. **FB 52**

Refurbishing stairs in dwellings to reduce the risk of falls and injuries. **FB 53**

Dealing with difficult demolition wastes. **FB 54**

Security glazing: is it all that it's cracked up to be? **FB 55**

The essential guide to retail lighting. **FB 56**

Environmental impact of metals. **FB 57**

Environmental impact of brick, stone and concrete. **FB 58**

Design of low-temperature domestic heating systems. **FB 59**

Performance of photovoltaic systems on non-domestic buildings. **FB 60**

Reducing thermal bridging at junctions when designing and installing solid wall insulation. **FB 61**

Housing in the UK. **FB 62**

Delivering sustainable buildings. **FB 63**

Quantifying the health benefits of the Decent Homes programme. **FB 64**

The cost of poor housing in London. **FB 65**

Environmental impact of windows. **FB 66**

Environmental impact of biomaterials and biomass. **FB 67**

DC isolators for photovoltaic systems. **FB 68**

Computational fluid dynamics in building design. **FB 69**

Design of durable concrete structures. **FB 70**

The age and construction of English homes. **FB 71**

A technical guide to district heating. **FB 72**

Changing energy behaviour in the workplace. **FB 73**

Lighting and health. **FB 74**

Building on fill: geotechnical aspects. 3rd edn. **FB 75**

Changing patterns in domestic energy use. **FB 76**

Embedded security: procuring an effective facility protective security system. **FB 77**

Performance of exemplar buildings in use: bridging the performance gap. **FB 78**

Designing out unintended consequences when applying solid wall insulation. **FB 79**

Applying Fabric First principles: complying with UK energy efficiency requirements. **FB 80**

The full cost of poor housing. **FB 81**

The cost-benefit to the NHS arising from preventative housing interventions. **FB 82**

Measuring fuel poverty. **FB 83**

For a complete list of IHS BRE Press publications visit www.brebookshop.com